# THE WEATHER AND A PLACE TO LIVE

WINNER OF THE
CENTER FOR DOCUMENTARY STUDIES/HONICKMAN
FIRST BOOK PRIZE IN PHOTOGRAPHY

# THE WEATHER AND A PLACE TO LIVE

## PHOTOGRAPHS OF THE SUBURBAN WEST

STEVEN B. SMITH

PUBLISHED BY DUKE UNIVERSITY PRESS IN ASSOCIATION
WITH CDS BOOKS AT THE CENTER FOR DOCUMENTARY STUDIES
DURHAM, NORTH CAROLINA

To Joe
Thanks

FOR MY FATHER

# CONTENTS

# INTRODUCTION
BY MARIA MORRIS HAMBOURG

Origins determine so much: they rule our tastes and politics as surely as they shape our noses. Just as no tree draws life except by its roots and no drawing exists without its page, none of us finds a way to maturity without reference to our background.

Artistic traditions are similarly pervasive. Even the supposedly untutored or "primitive" dauber likely had his eye on some eccentric precedent, while most artists, by the time they are in full command of their craft, have consciously collected their kit of exemplary sources and solutions. These personal pantheons usually mix the idiosyncratic and the more broadly generational, and the resulting mélange constitutes that individual's usable artistic tradition.

Steven B. Smith photographs in the American West with a large-format camera, producing landscapes that seem to this viewer to be in the direct lineage of the nineteenth-century pioneers who photographed the same territories with a similar breadth, formal elegance, and concision. Surprisingly, however, Timothy O'Sullivan and Carleton Watkins served not as inspiration for Smith, but as confirmation of his choices of subject and style. He actually entered the stream of photographic tradition farther downriver.

Born and raised near Provo, Utah, Smith enrolled at the state university in 1981 with the solemn ambition of becoming a painter. Having used a camera as a journalist's tool during high school, he did not consider photography a calling equal to painting, but he changed his opinion around 1983 when he came upon Robert Frank's book *The Americans* (1959). From it Smith learned how individual photographs could form a serious and sustained sequence, "something like a long poem," with themes and variations weaving a complex polyphony of enduring resonance. As a Swiss citizen, Frank had witnessed urban and rural scenes new to his experience on a road trip across America and his outsider's perspective afforded him telling insights into the cultural and social condition of the nation at mid-century. Clear-eyed and stark yet lyrical, and sweeping the whole nation with indictment, Frank's great book seduced Steven Smith, as it has so many smart and sensitive youths who observe and question the status quo. Yet the challenge of *The Americans* was daunting: how could a stripling planted in Utah learn to grasp, as Frank did so masterfully, the potency of the prosaic, to find and frame simple truths hiding in plain sight?

If Frank was hard to emulate it was somewhat easier to follow Garry Winogrand, another remarkable photographer, a virtual whirlwind with a Leica, whose athletic, unruly style and modern subjects – airports, press ops, and the like – brought Frank's vision up to date. Winogrand was a "god" to Smith's generation, and as long as one was focused on the social world his sprawling body of work was an inexhaustible resource. Even as Smith photographed the college students and street trawlers in Salt Lake City, however, he was beginning to realize that it was not how people looked or interacted but their deeper cumulative effect on their environment that moved him. He sensed that he could make an extended portrait of society by attending to the values unconsciously revealed in its physical surroundings. This was confirmed when he discovered Robert Adams's *The New West* (1974), an ideal essay on the beauty and desecration of Colorado, limpid photographs of the land and the light, the tract houses, highways, mountains, and motels. Adams had uncovered the shape of a new artistic terrain and mapped it so deftly that his work, as a model, was moot: To appropriate that simple, quiet eloquence as a *style* was to empty it of meaning, and meaning was what Smith was after.

Sometime before he graduated Smith also came across Lewis Baltz's *Park City* (1980), which documents the crash transformation of a sleepy, former mining town in northern Utah into a major ski resort. He thought the transparent, forensic views and the rapid "Californization" of the natural site both equally "ugly." The pictures were affectless and the subject depressing; maybe this wasn't even art? The tide of cheap houses overrunning the land near home Smith knew only too well, for each time he had to drop out of college to earn money to continue his education he worked on construction crews building newer versions of the very same cookie-cutter houses depicted in *Park City.* The book rankled and stuck with him.

In 1987 Smith moved to Connecticut to attend graduate school at Yale where he ran head-on into a mountain of photographic tradition. The classic canon as conveyed at Yale was built on the shrine of Walker Evans, who had taught at the school until his death in 1975; it reached back to the turn-of-the-century French documentarian Eugène Atget, and yet farther back to the most poetic and reductive photographer of the Civil War and western frontier, Timothy O'Sullivan. Together with Atget and O'Sullivan, Evans was the champion of the intelligent photographic report in which beauty is but a byproduct of cogent and sensitive description. Starting with straight, astringent depictions of the relevant evidence of a culture, a life, or a landscape, these photographers built up catalogs of images that tracked underlying conditions and conveyed the sorts of large, amorphous cultural truths that are otherwise difficult to peg down. The tradition was alive in many contemporaries: besides Richard Benson and the current crop of Yale professors, most important to Smith were Robert Adams, Lewis Baltz, Joe Deal, Frank Gohlke, and others who had been shown together in an exhibition in 1975 called *New Topographics.* These young artists used large-format cameras and black-and-white film to describe the American landscape in great detail and without apology or celebration. Their pictures acknowledged that the romantic, historic idea of the natural landscape was no longer even a usable myth; the reality, at least in all urban and many suburban sites, was man-made environment, prefabricated and commercialized.

On relocating to Los Angeles in 1991, Steve Smith found his vision changed. With new eyes from his four years in the Northeast he now recognized the magnitude of the changes in the western landscape as phenomenal and cataclysmic. Abetted by his firsthand knowledge of construction acquired as contractor for his own company in L.A., and dividing his time between construction sites and road trips from southern California to various locales in Utah, this native son was perfectly positioned to apprehend how the desert in southern California, Nevada, Colorado, and Utah was metamorphing into a vast suburban frontier. He was aware that he was confronting another chapter of the "ugly" from which he had recoiled in *Park City*, but without the confusing filter of Baltz's aesthetic and with the confidence of his own gut reaction, he now experienced a tantalizing ambivalence. Surely what was happening was folly masquerading as progress, capitalist venality defiling the land . . . yet these developments had an unmistakable scope and hubris, a shadow even of the grandeur inherent in the large-scale earthworks of Robert Smithson and Michael Heizer. Just as their installations stood in dramatic relation to the ancient planetary origins of the land, so the optimistic, new housing developments, in their very evident recency, pointed up the primordial nature of this geology. Smith was riveted not only by these connotations but also by what he was seeing as patterns of stratification, sedimentary layers such as an archaeologist from another planet might study.

The boom in real estate sent out waves of housing that first rolled out across the easy valleys, later jumped up to the view-crowned summits, and lastly climbed back down the lower, less-desirable slopes to fill in toward the center. When the available land was completely covered and the angular arroyos were lined with cement, the boom was done. Poet James McMichael observed the same commercial and geographic devolution in the Pasadena area earlier in the twentieth century. Of this process, in *Four Good Things* (1980), McMichael concluded dryly: "What they were selling was the weather and a place to live, and that was what was left."

Smith's photographs of the metamorphosis reveal a discerning knowledge of the processes – the laying out of subdivisions and roads; the necessary sculpting, terracing, retaining, and shoring-up of the earth with boulders, berms, and cement moldings; the careful channeling and managing of run-off water; the fastening of topsoil with miles of jute and rolled fiber brakes; the replanting of thousands of identical seedlings, bushes, and groundcovers. These activities would be easy to interpret as the imposition of unwanted control, a subjugation of nature's plan to man's will, and the concrete culverts, scraped hillsides, and sprinkler systems would be damning evidence in a hypothetical trial of the developers. Conservationists would decry the destruction of natural habitats, the myopia of profligate water use, and the wrong-headed arrogance of commercial exploitation, while native poets might plead for the inviolability of their sacred sites. And the photographer would take their side.

As a boy Smith lived two small-town blocks from the expansive, open fields at the base of the mountains where he ran with his friends, hunting birds and playing in dug-out hideaways. Like most of us, he prefers the land in its prelapsarian state yet recognizes that his regret for the landscape of his youth is rooted in personal nostalgia similar to the poignancy anyone feels in the loss of a childhood home. But Smith's lament is keener and broader than the usual, and his dismay is laced with other sentiments, not the least of which is the operative irony governing his activity: what he finds by turns deplorable, grand, and ludicrous is also what butters his bread and buys his film. The photographs reveal his conflicting feelings; some are bleak, some seem born of disbelief, others convey a sense of awe, and many are both funny and sad at the same time. That there is ambivalence is not surprising, for how would we honestly regard this scene if we lived in the immediate area or benefited from these developments?

Smith could have recorded a failure of the imagination or the ruin of desert ecologies but he was after something much more interesting and amorphous – an intersection of human, climatic, and geographic realms as yet without a name. Such an orderly, labor-intensive, wide-ranging application of knowledge and engineering to the land might be considered some novel and rampant form of *garden* if houses and streets were not its principal rationale, but since they are, this collocation is usually termed a suburb or a subdivision. Surely these are inadequate terms for Smith's subject, which, in its totality, is a vision of the future of our planet, of the time when man-made environments no longer just spread out in widening circles around cities and encroach like weeds along the highways, but hold sway everywhere, carpeting the land from valley to mountain and from sea to sea.

Concretions of cultural beliefs and practices, these landscapes are material evidence of who we are. They reveal a vision of the earth as property to be owned, physically shaped, delimited with boundaries, and viewed. Looked at from a distance, as Smith's photographs allow, the little parcels of private property appear empty, static, and sacrosanct, if occasionally endearing and whimsical. Wholly dependant on the delivery of an unlimited supply of water, electricity, and gas, some of these domiciles may boast plots of irrigated grass and heated driveways, but all of them have "conditioned" air and every other electronic and mechanical "necessity" of modern life. The spirit of the place is one of isolation, regularity, and sterility, above all of access and commodity – of going and coming and owning, but not of community.

What these photographs suggest about cur relation to the land is distressing. The houses are not attached to the land as, say, a farmhouse is; they are not the matrix of the landscape around them but the occupying force. The more dramatic the view or the stone outcropping, the more rocks in the rock garden, side yard, or retaining wall, the greater the self-satisfaction. The earth itself is material to be managed – scraped off, scooped out, infilled, and graded, but always it ends up a substrate to something else – the bec of a road or the slab of a house. It can be pasted over with concrete or blacktop strewn with boulders, or spread with pebbles or rolls of turf, but it has no real use except as the support of decoration and visual display. This land is not home to anything; the jackrabbits, gophers, scorpions, and even the birds are gone. What is in the pictures is what is left, the weather and a place to live.

# THE WEATHER AND A PLACE TO LIVE

They all made up a neighborhood, were part of that
difference from house to house that showed
more clearly in the mornings when the light
canted around the mountain, lifted, stayed . . .

The mountain . . . has severe
and angular back canyons where the light is always
unexpected, out of place, too simple for the
clutter of the granite blocks along the creeks.
The slopes have low rough shrubs, some firebreaks.
It rains sometimes, and then the soils wash easily
through . . . canyons to the small
catch-basins and the storage tanks. The bedrocks
tilt toward the west, and so the seepage
drains that way. . . .

What they'd been
selling was the weather and a place to live, and
that was what was left.

**JAMES MCMICHAEL, from *Four Good Things***

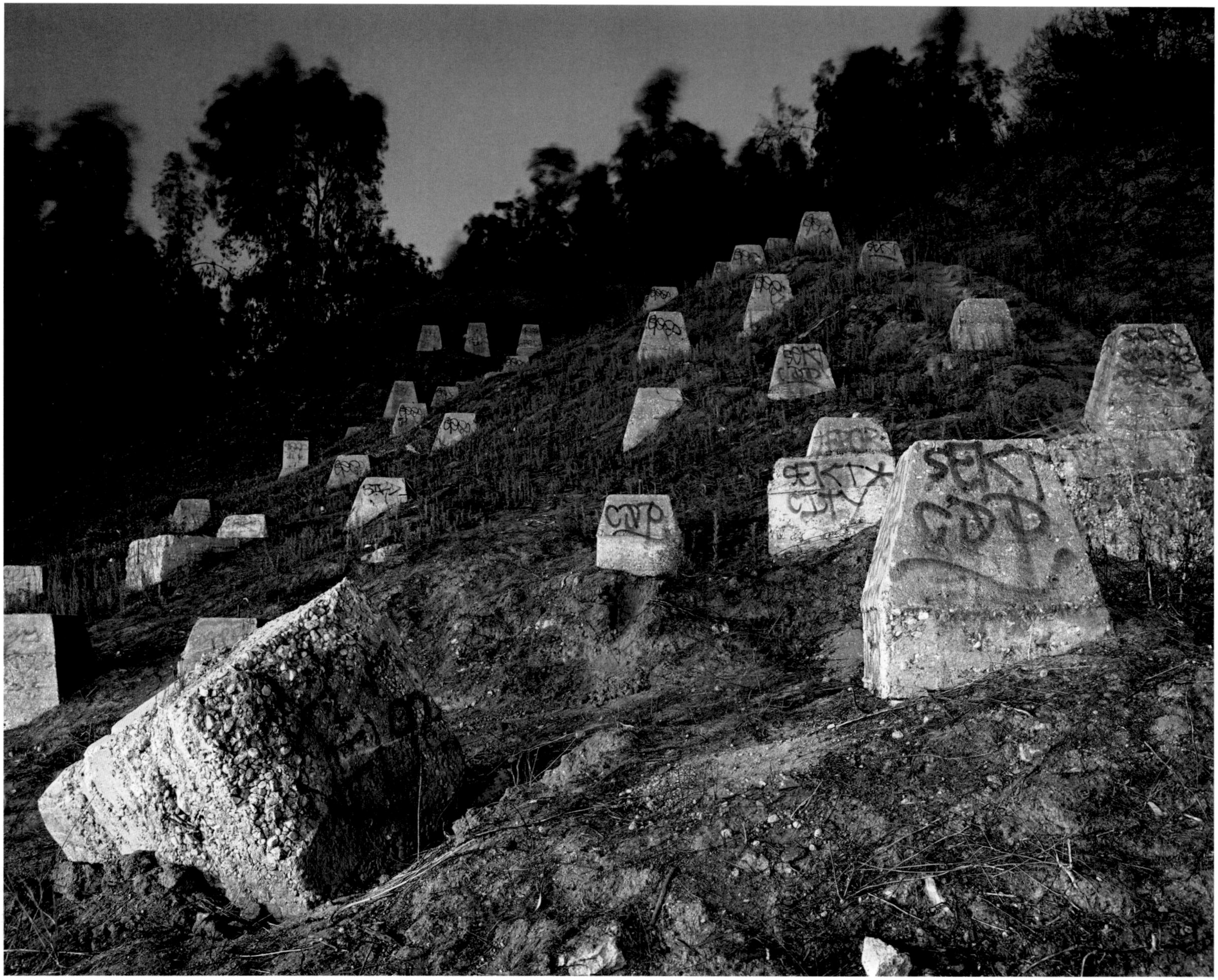

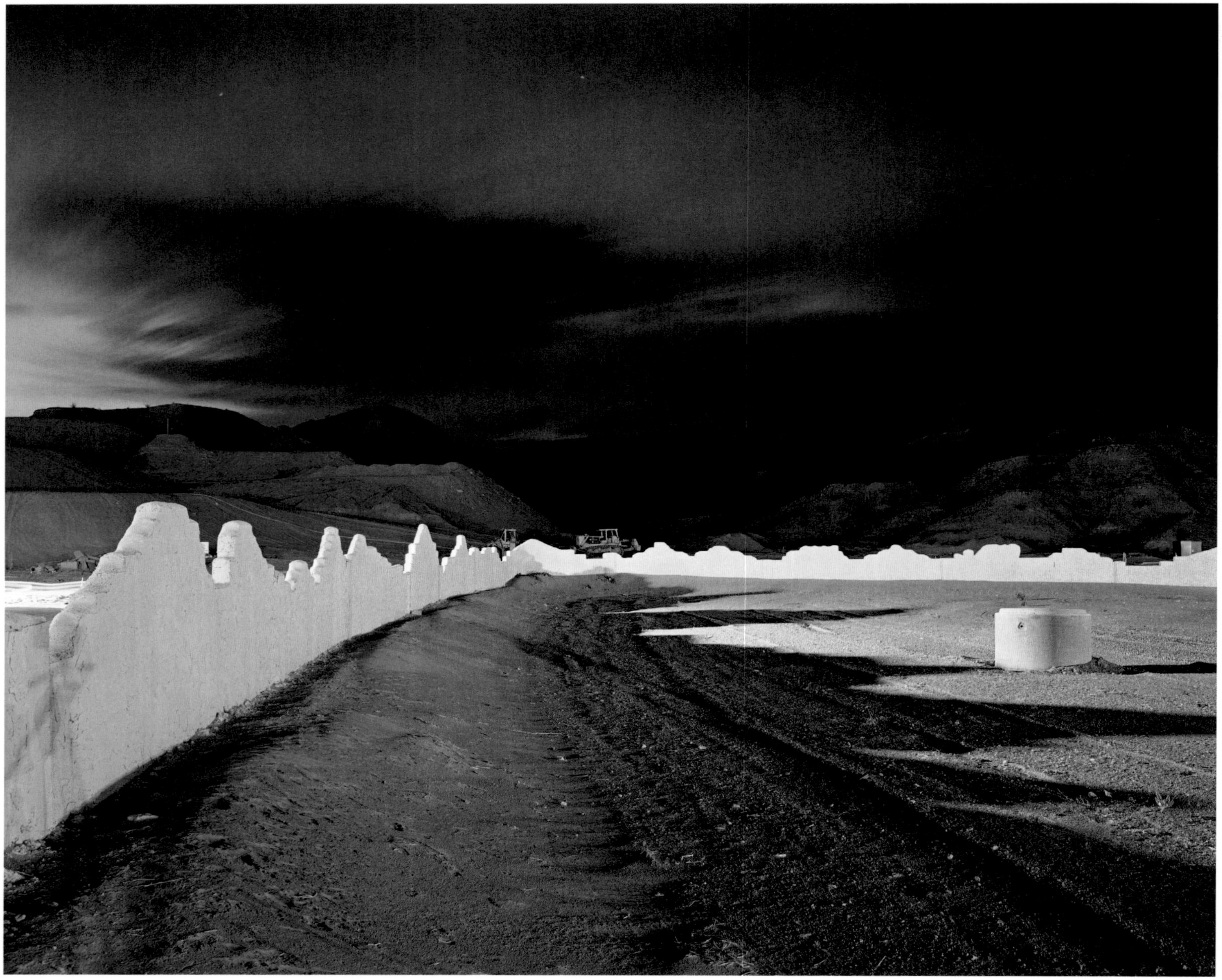

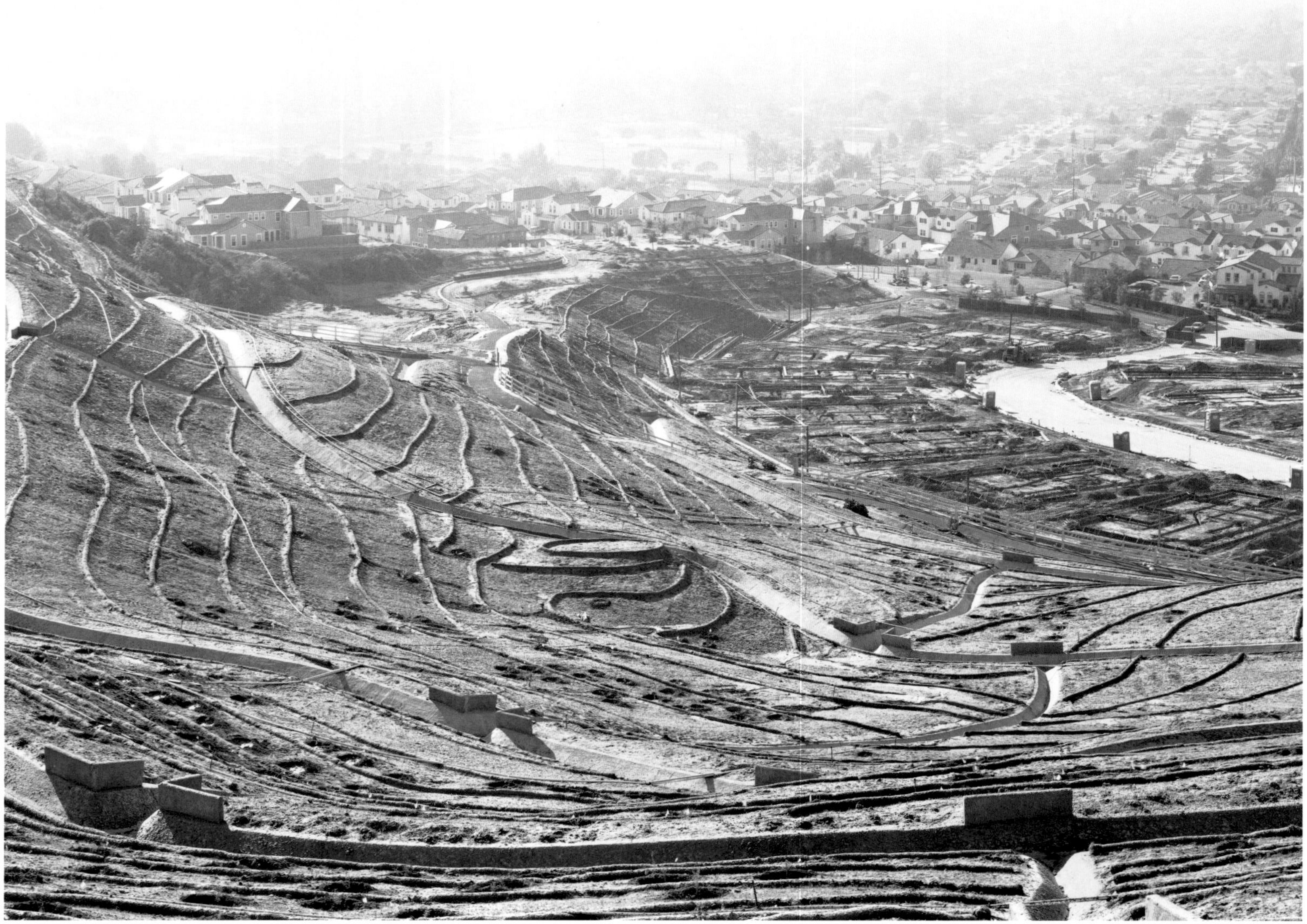

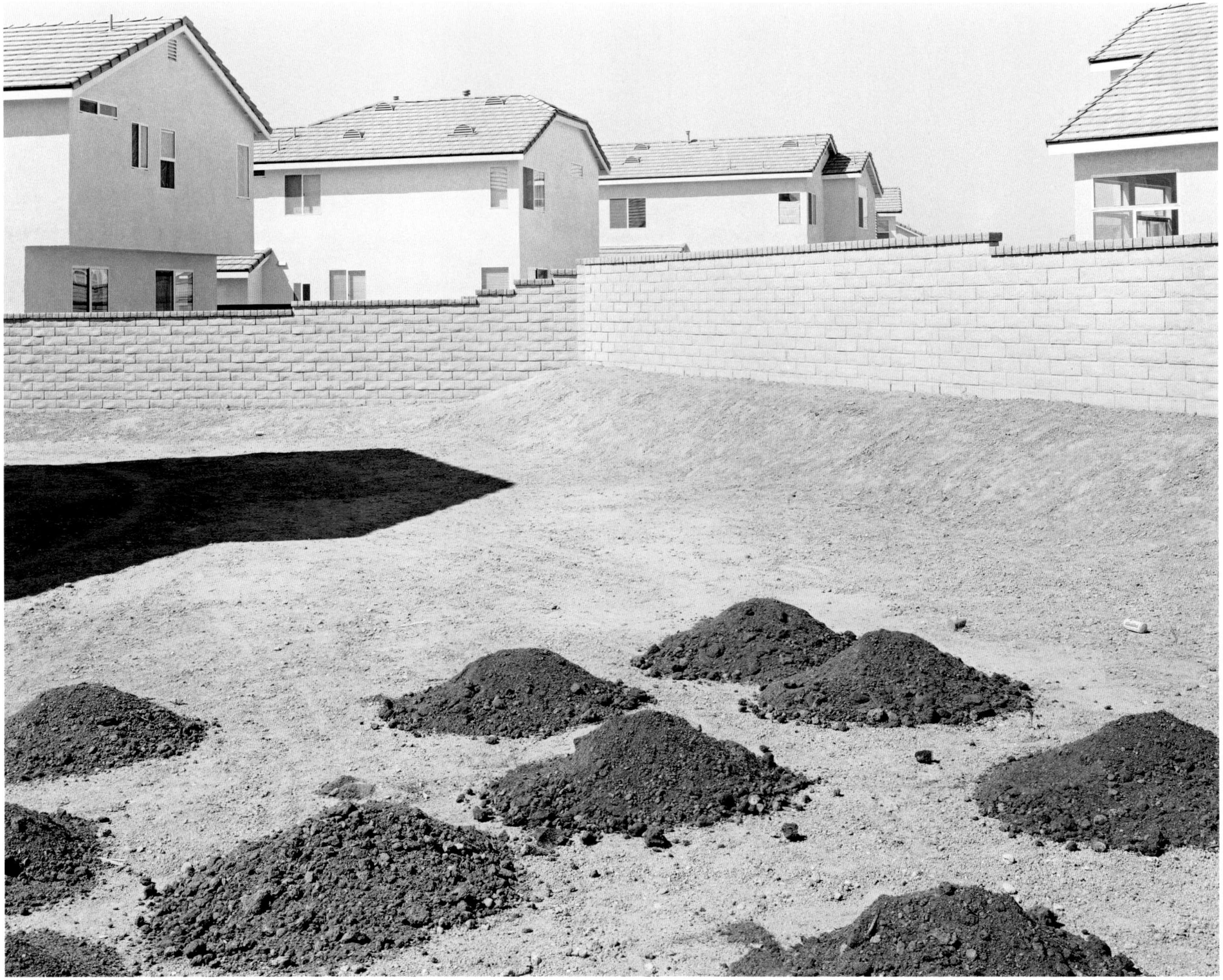

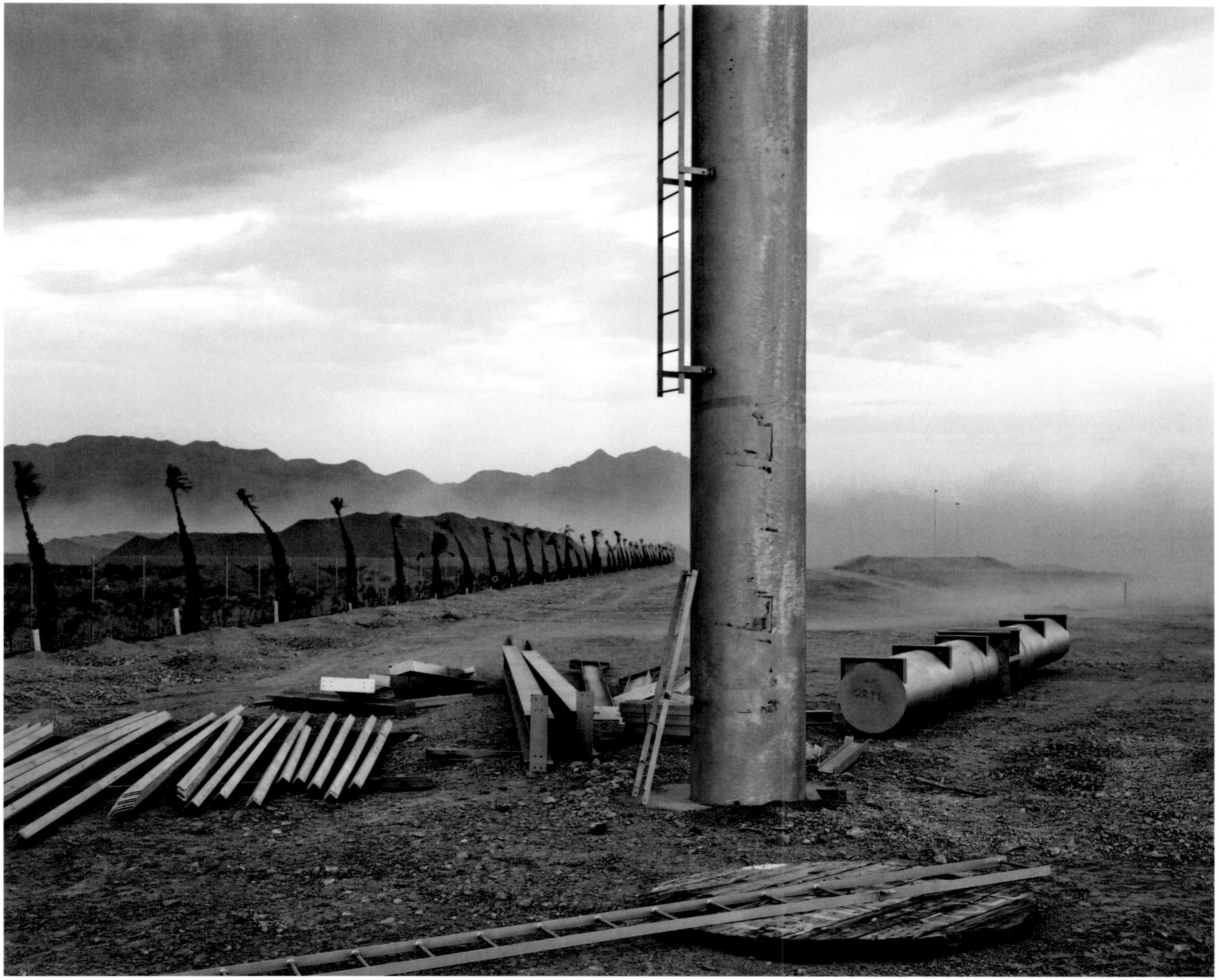

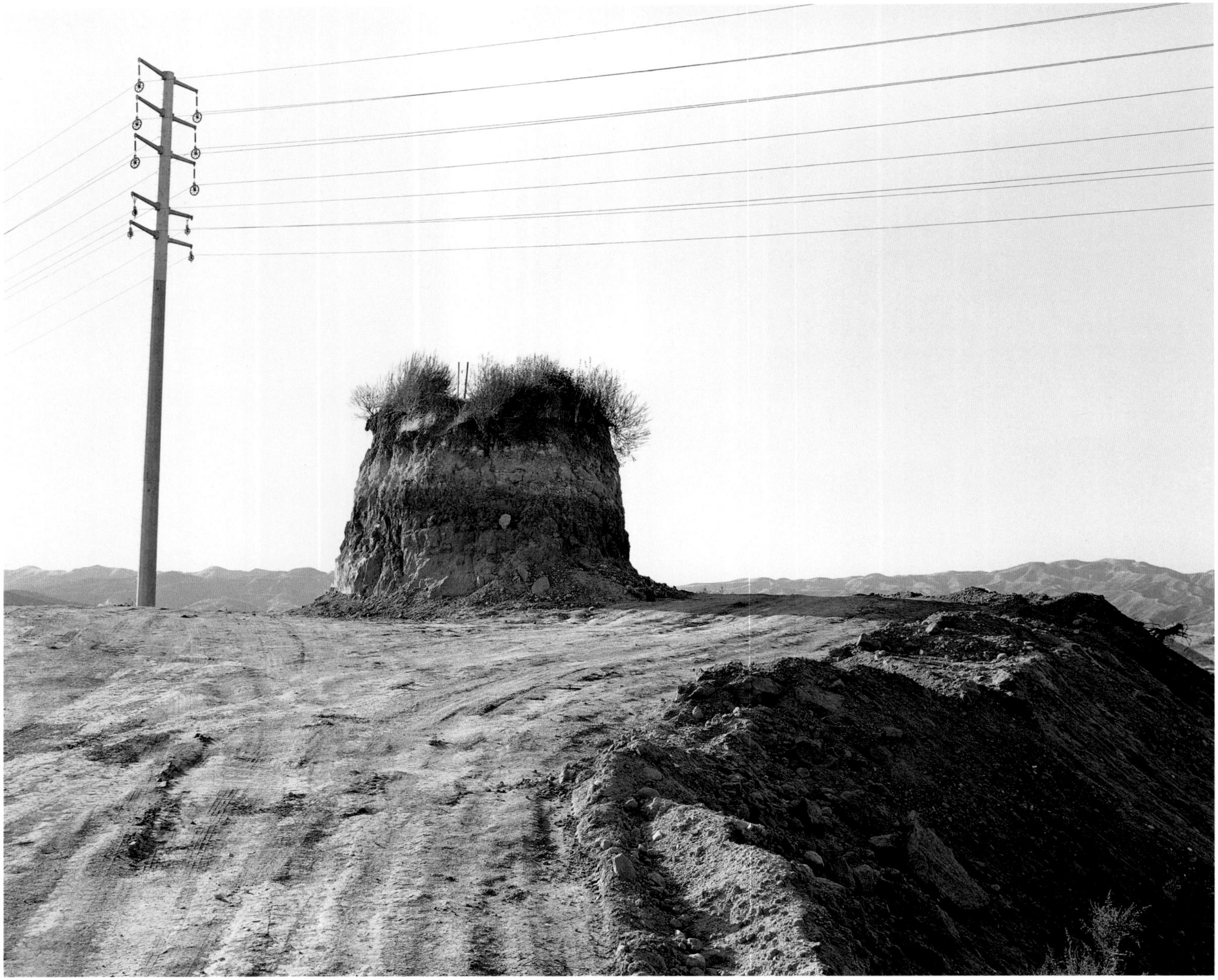

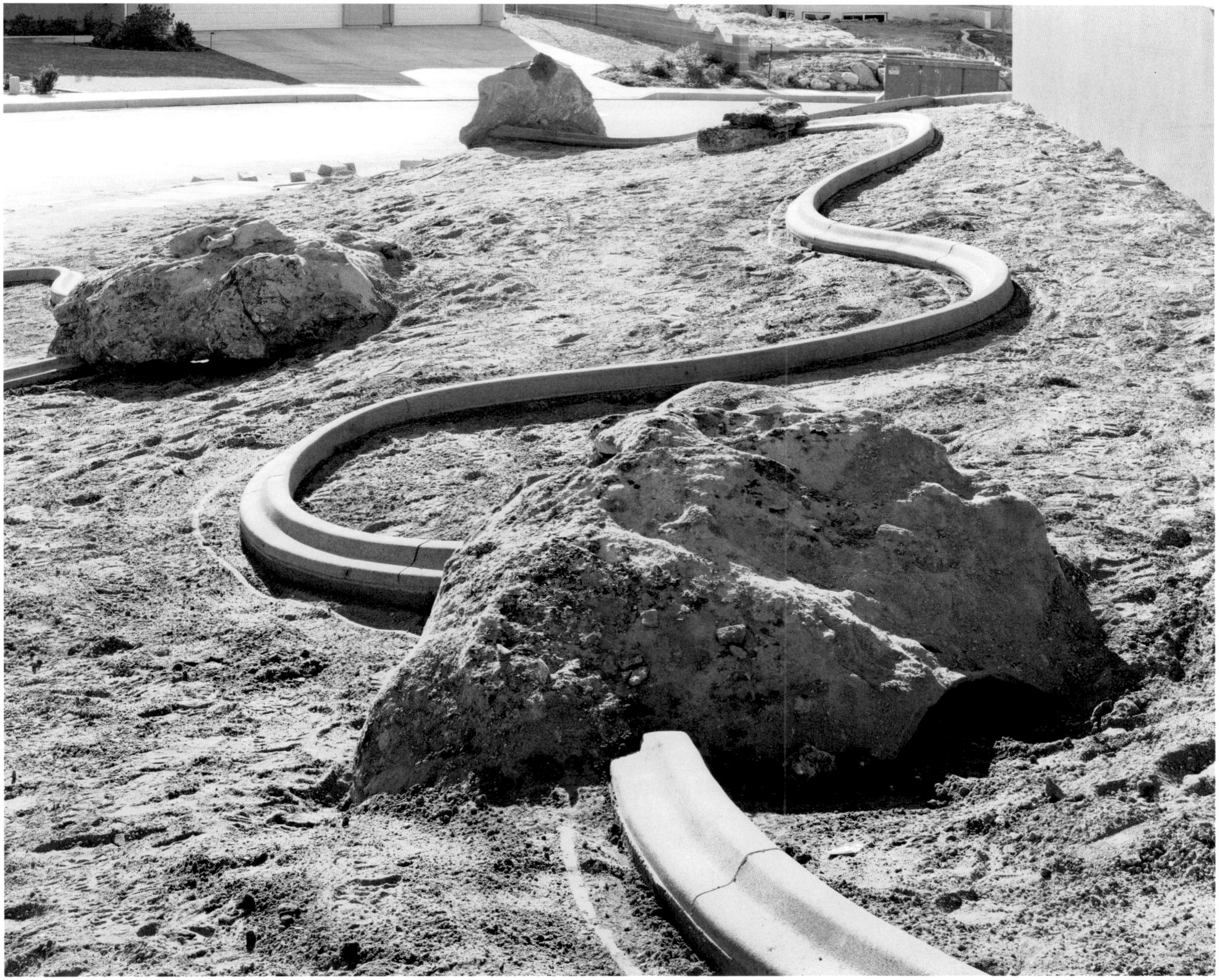

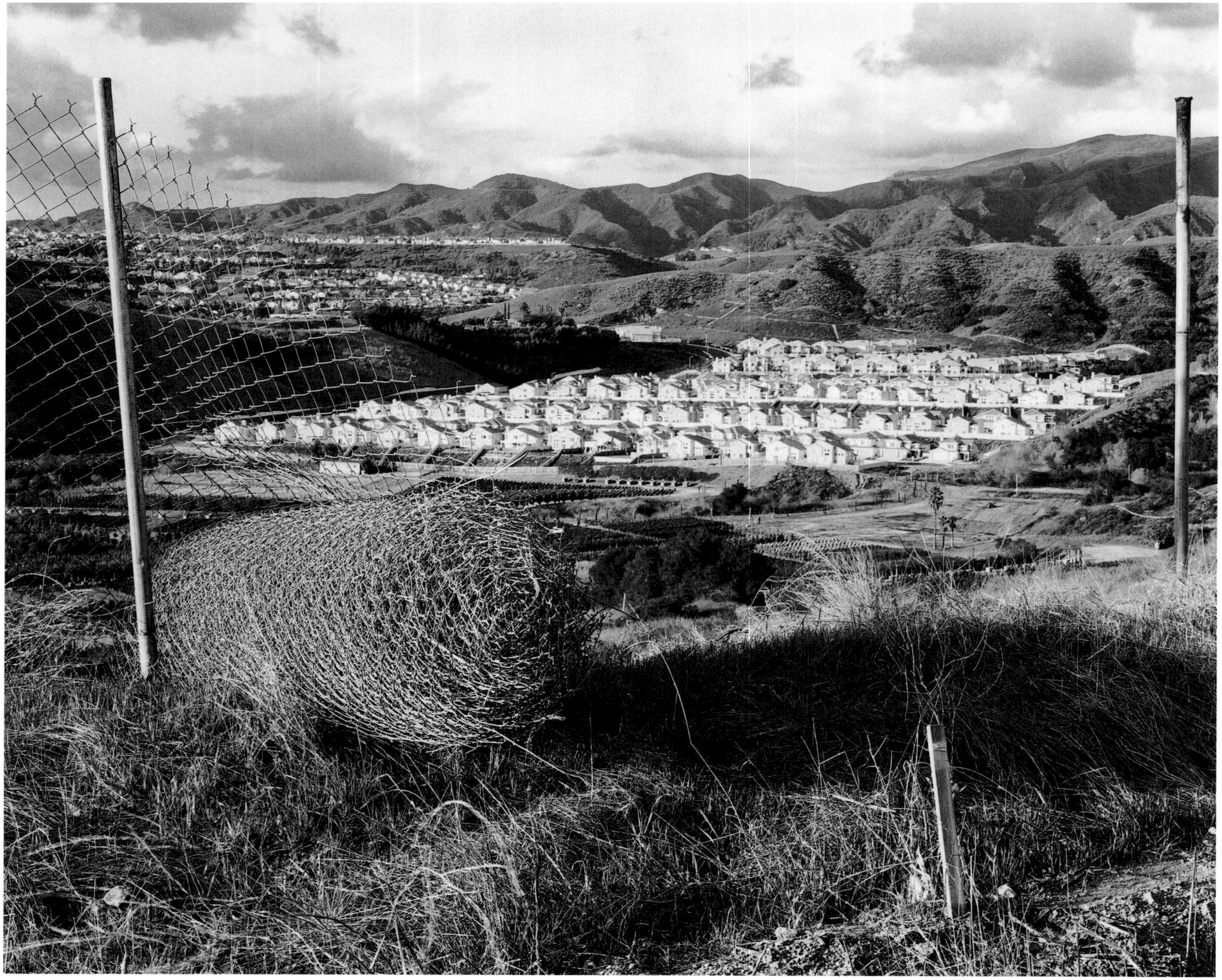

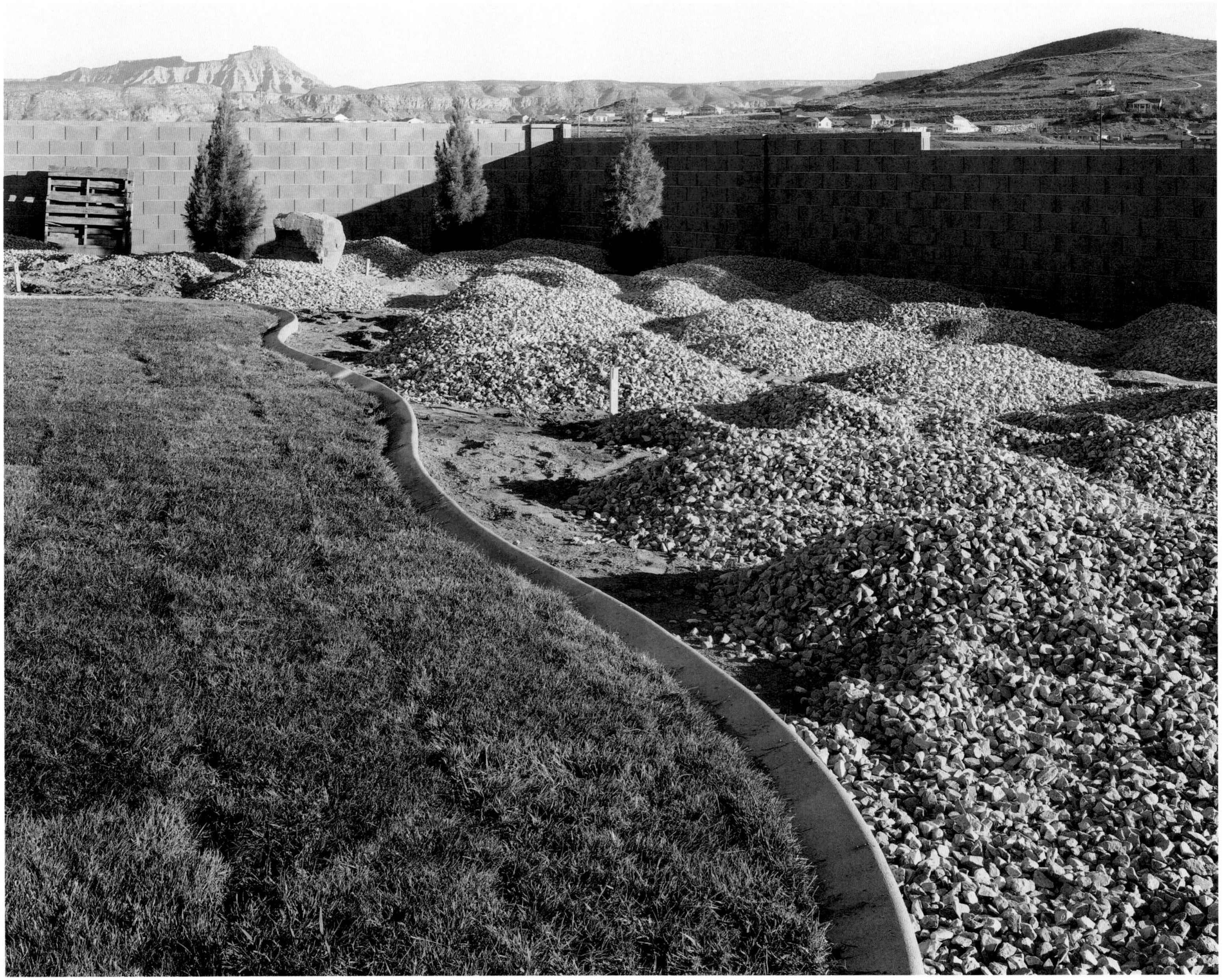

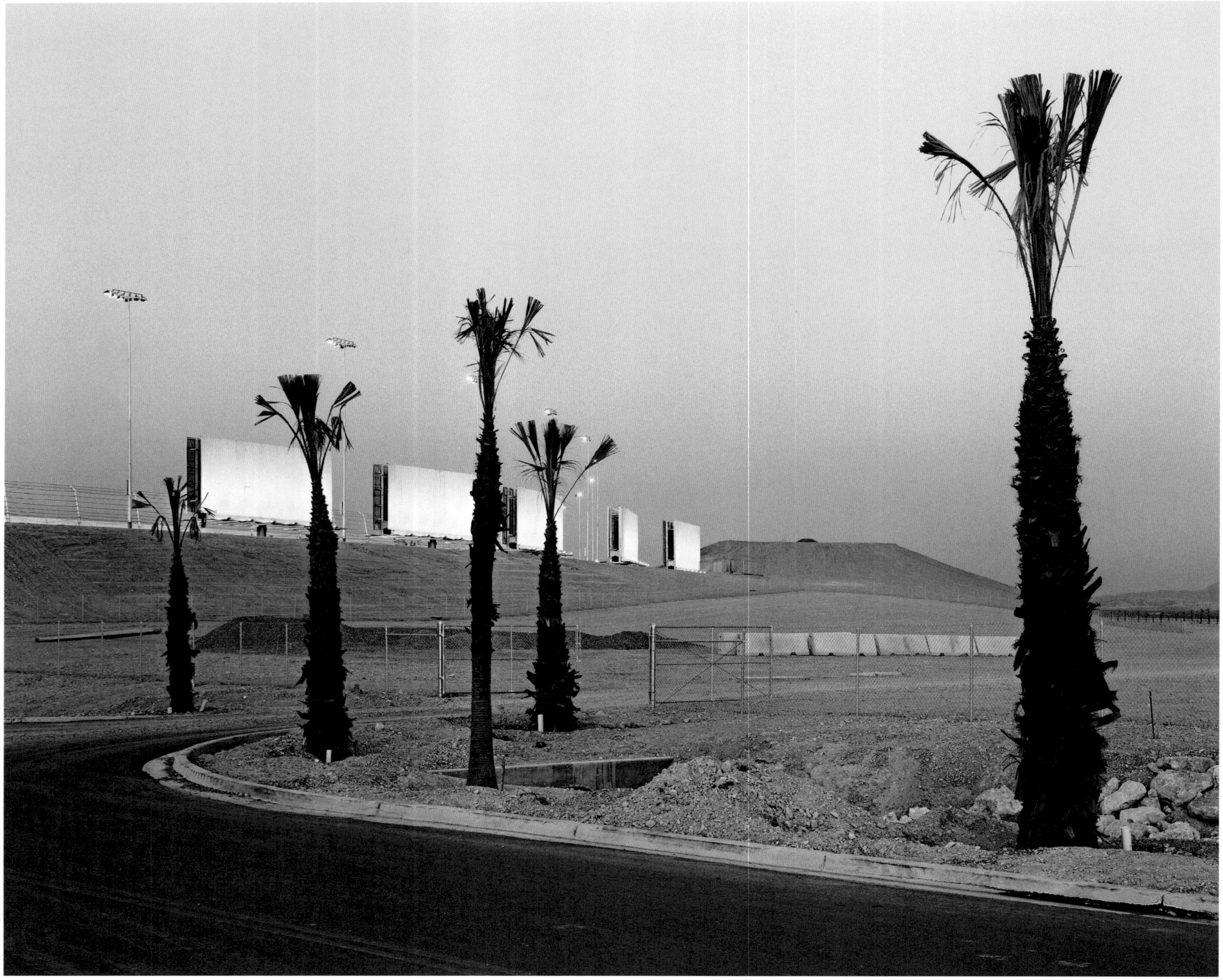

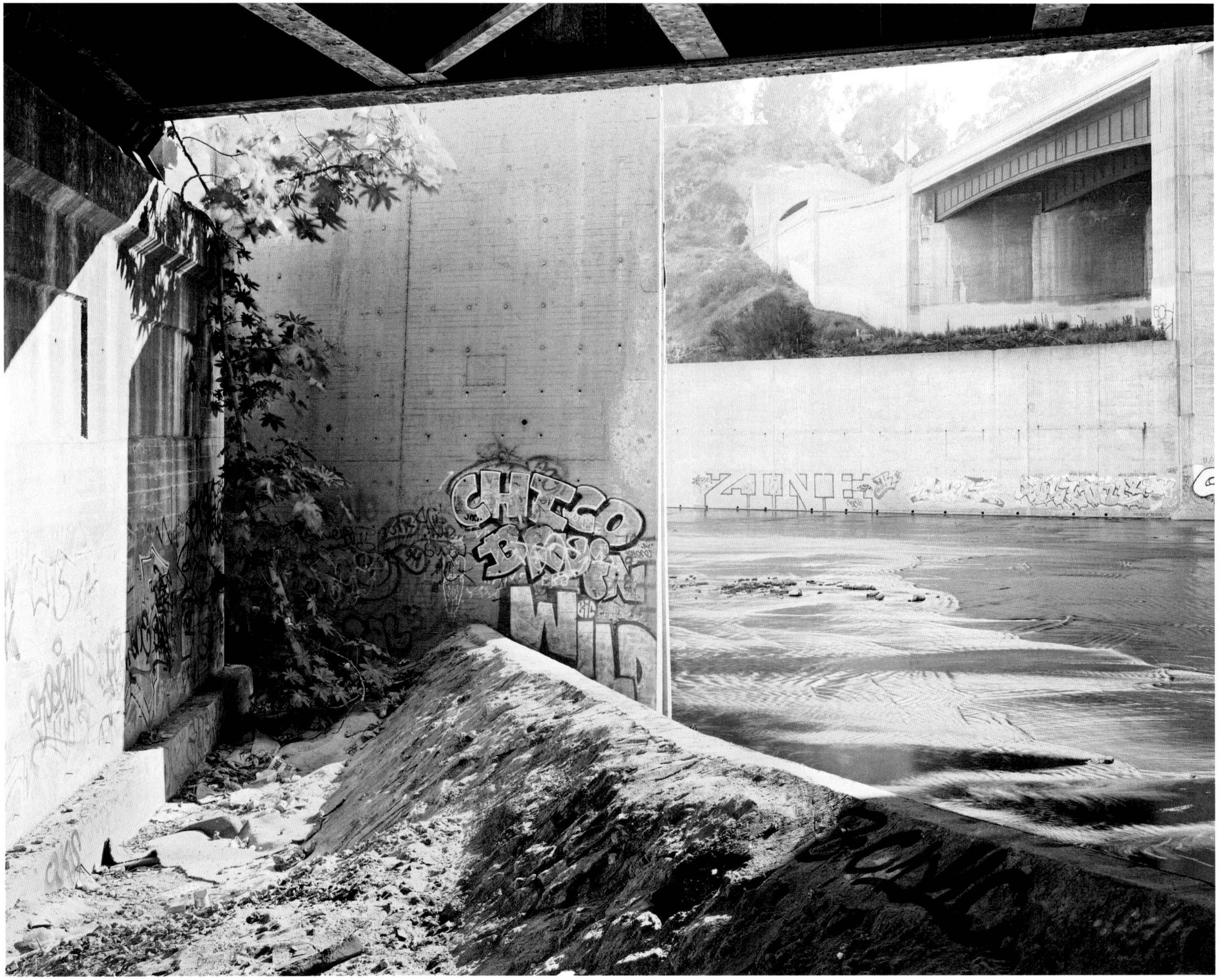
CHICO
WILD

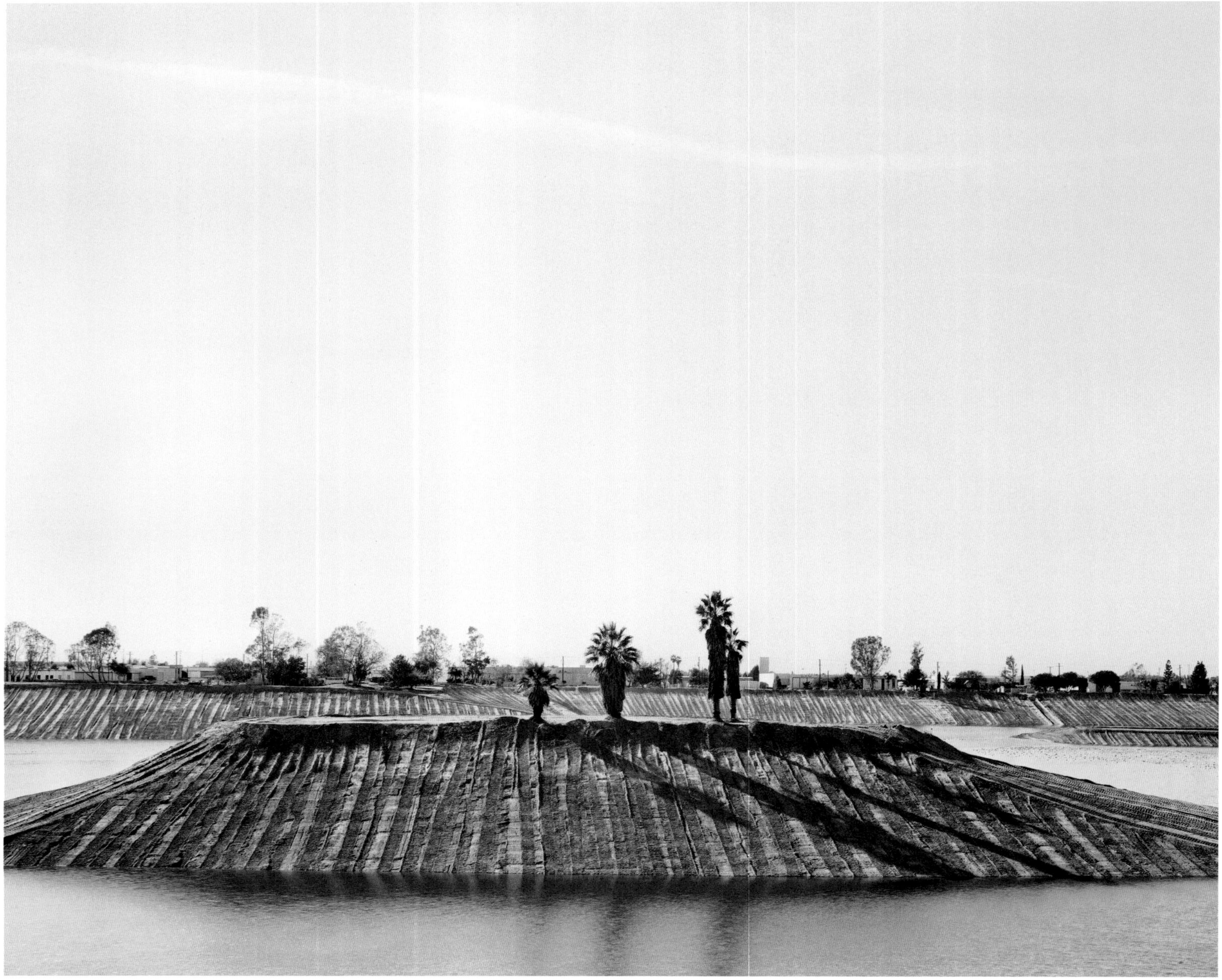

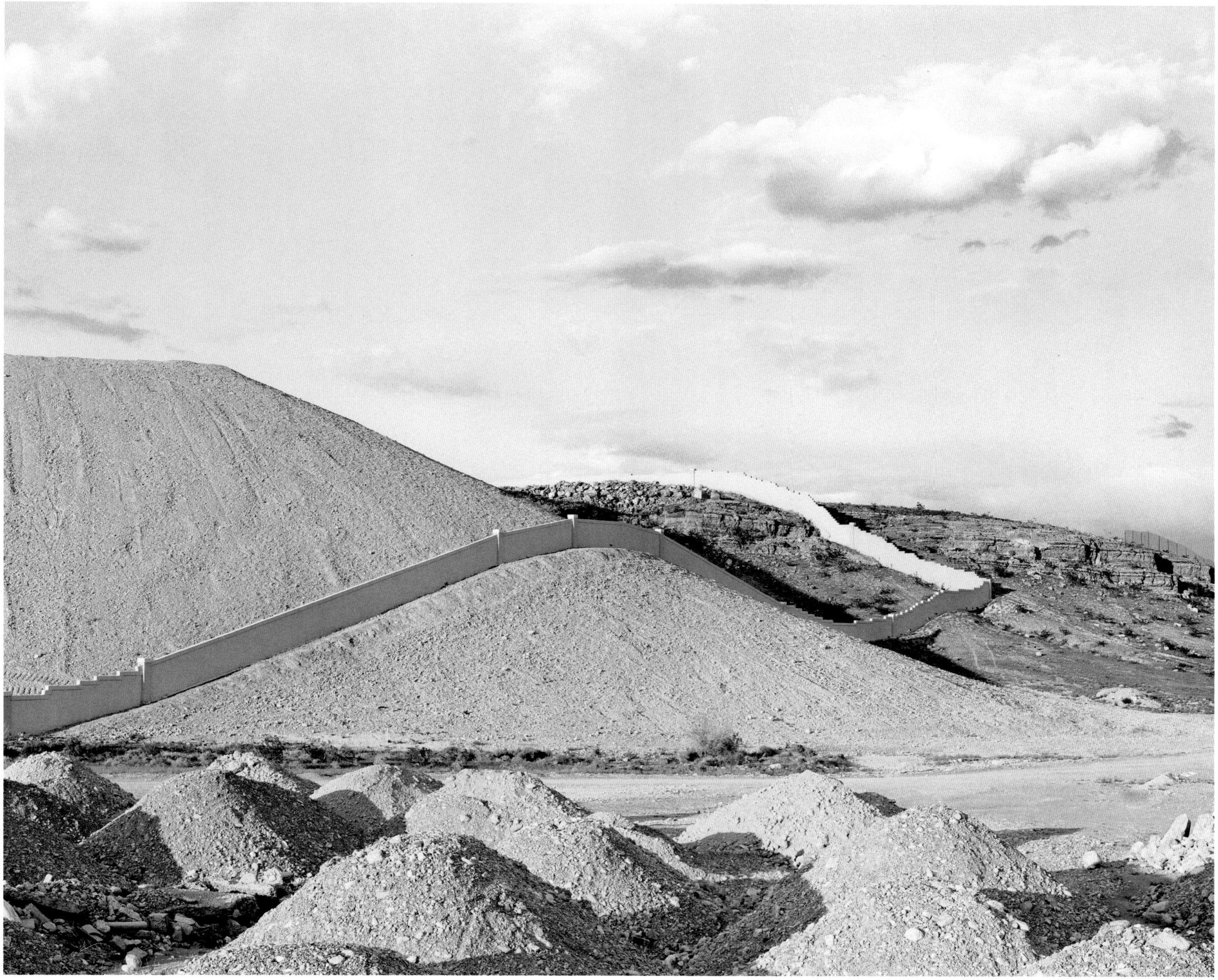

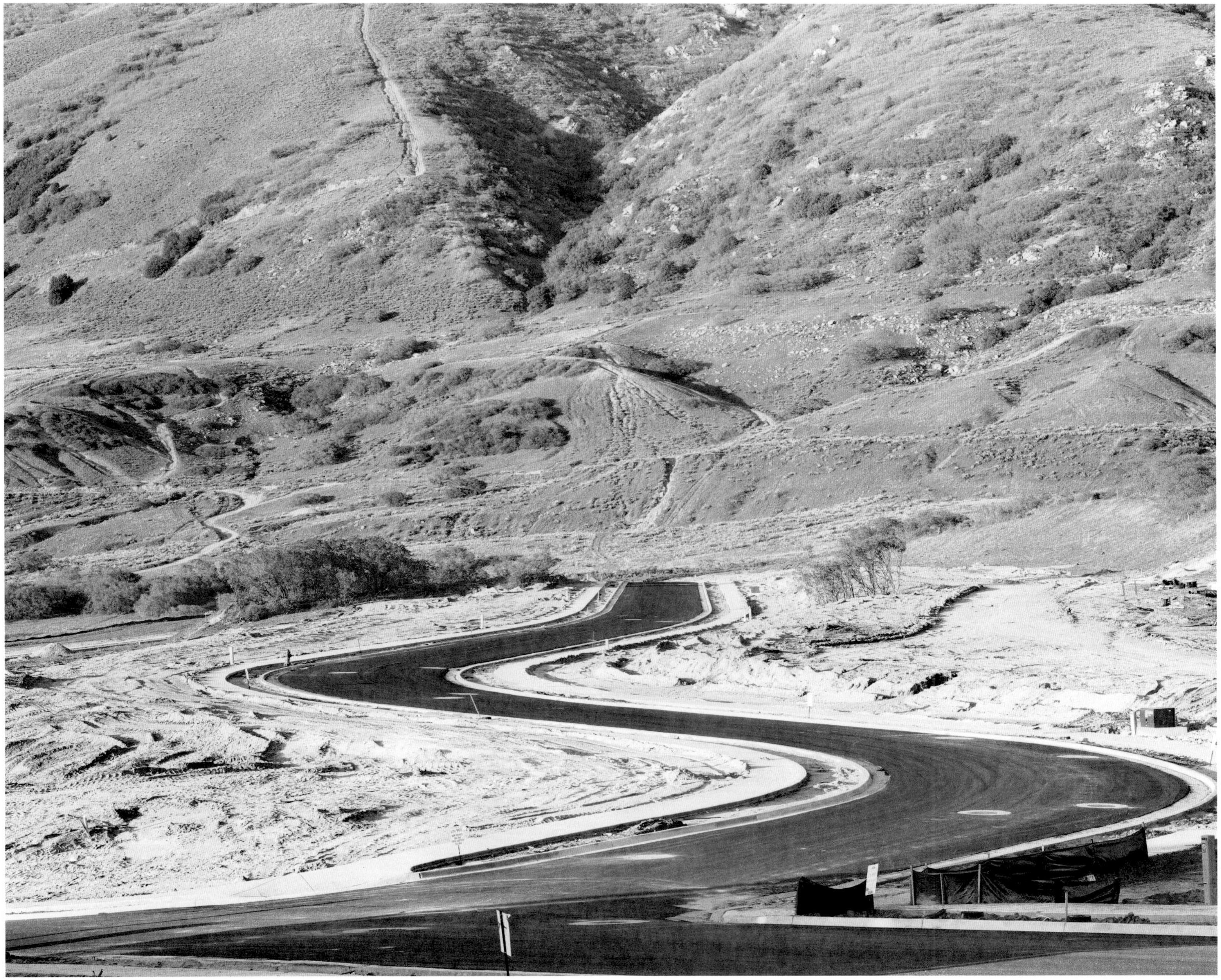

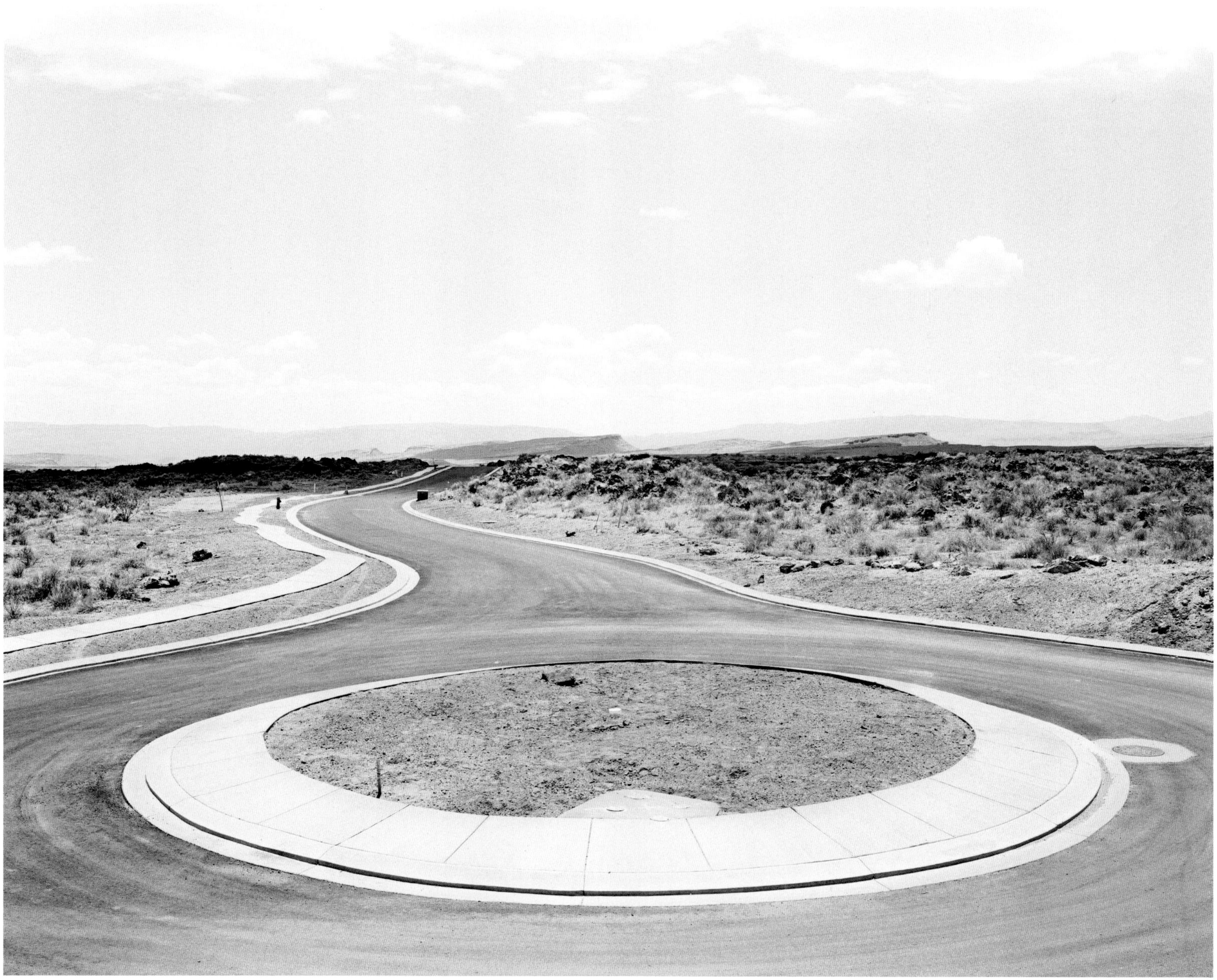

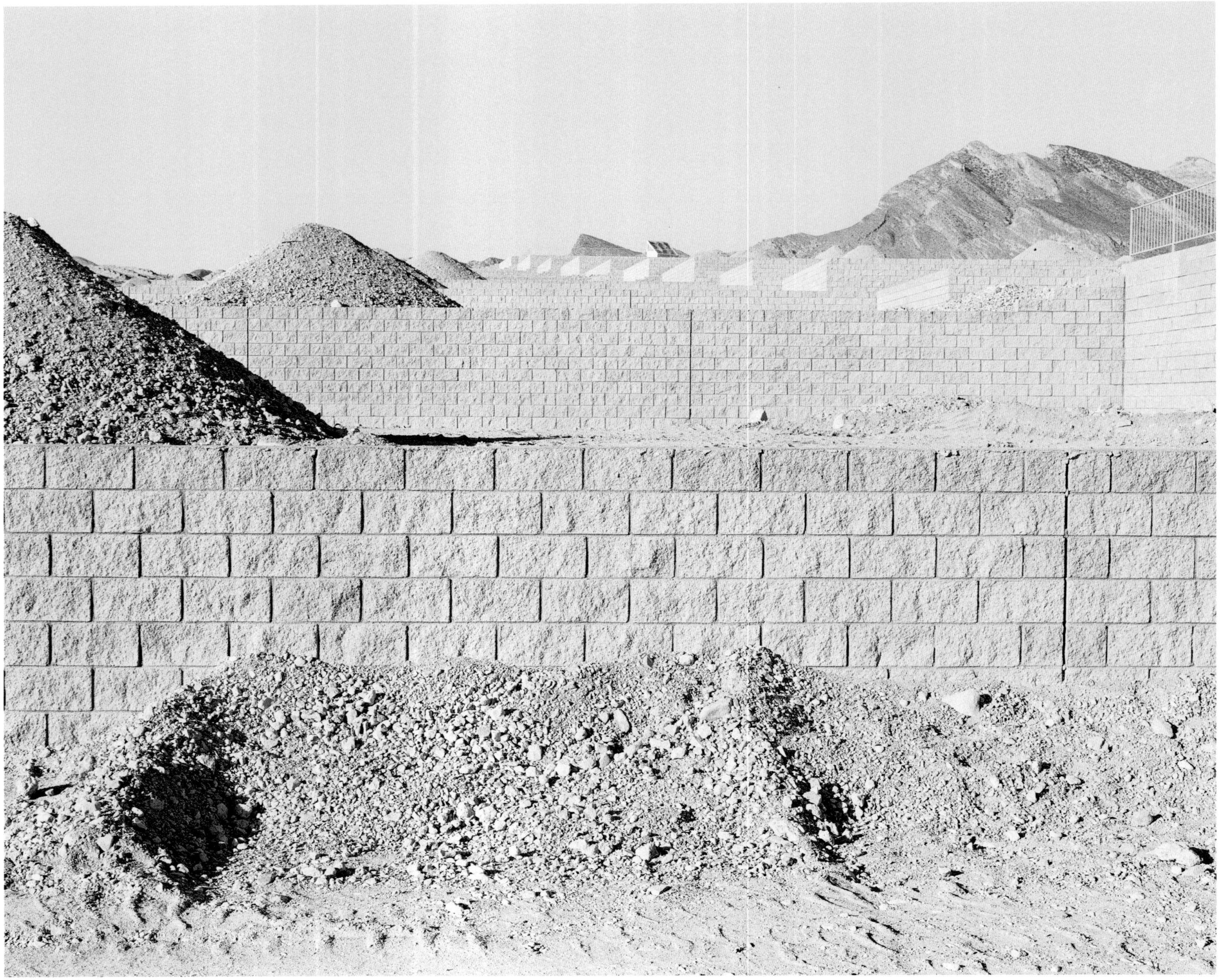

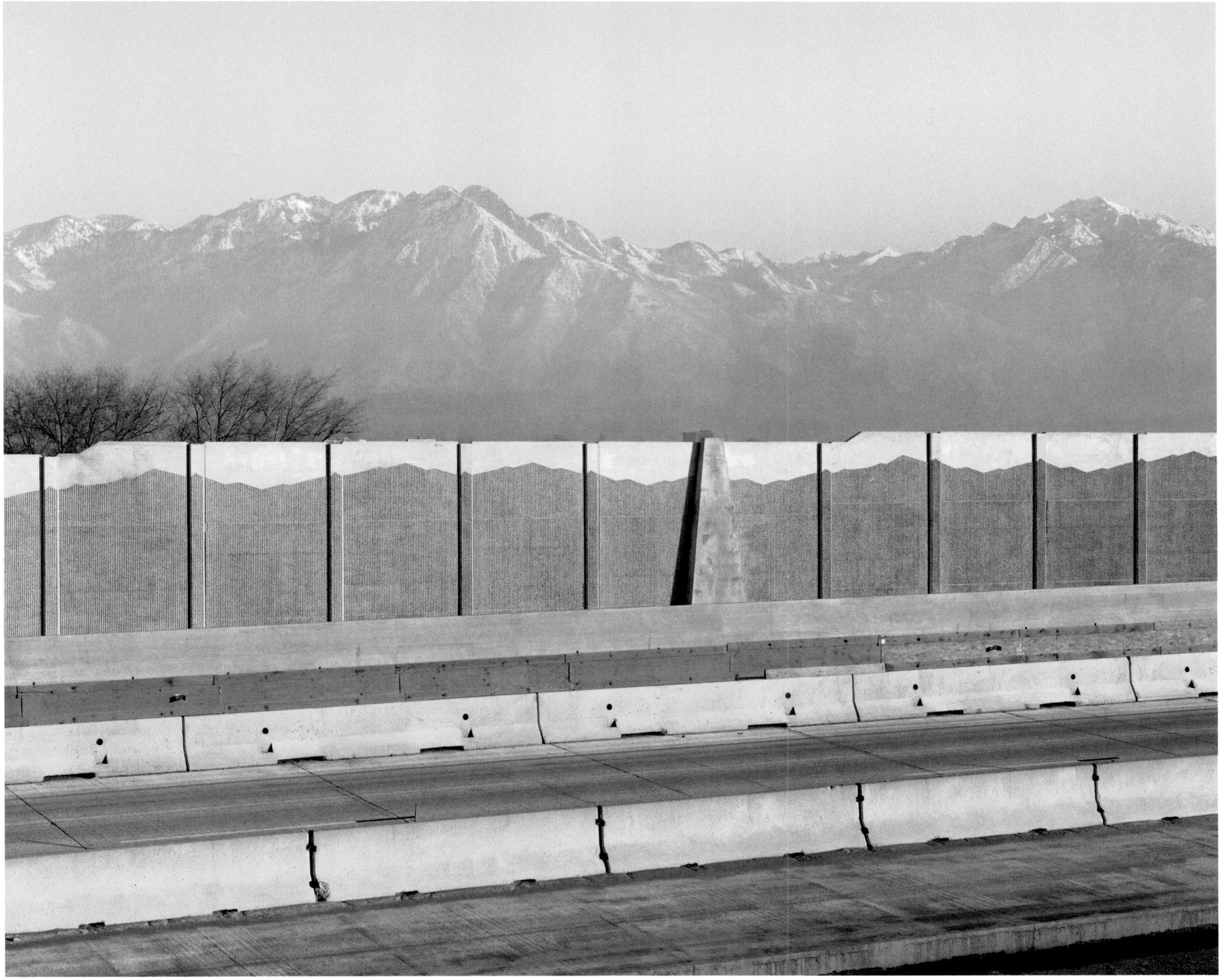

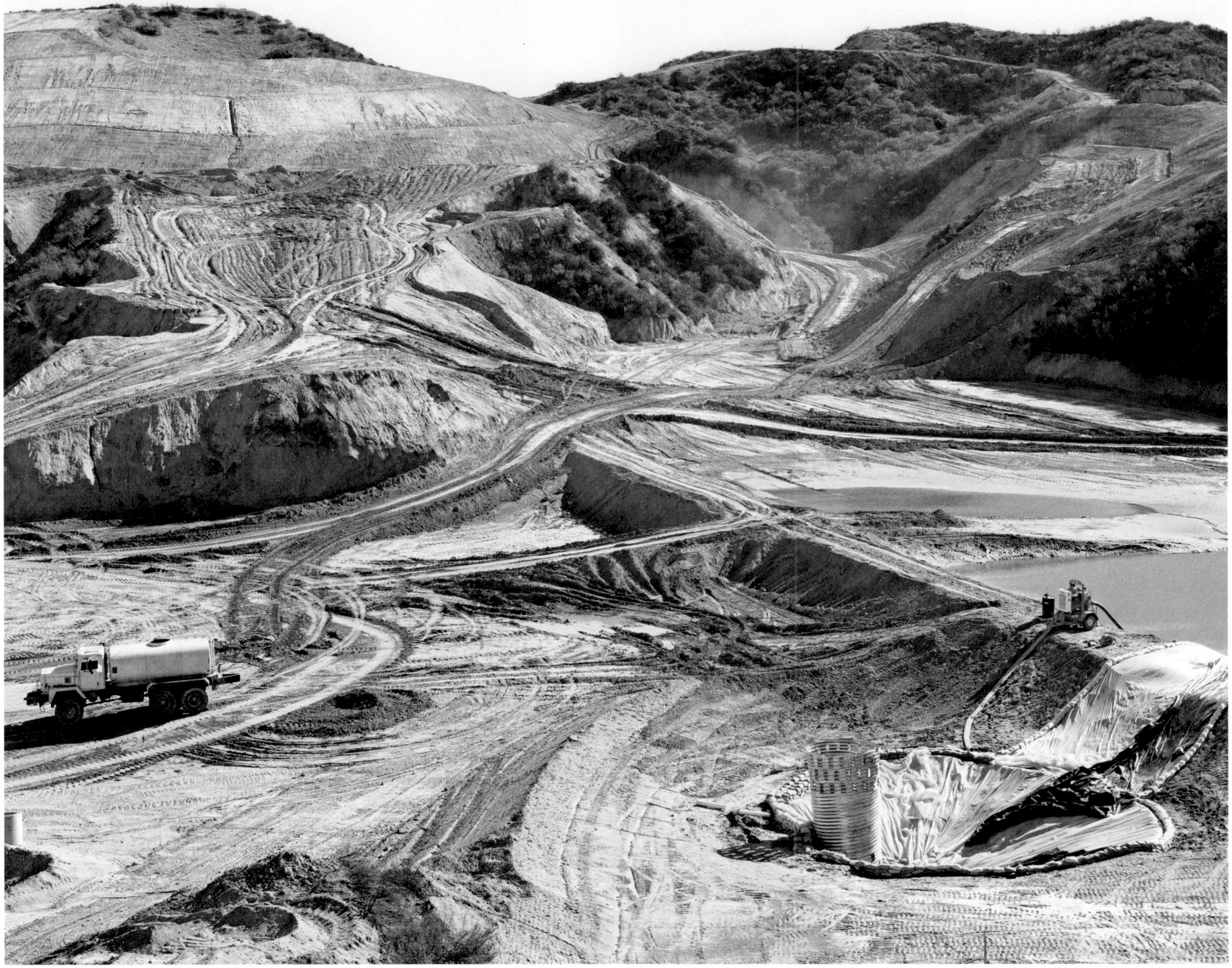

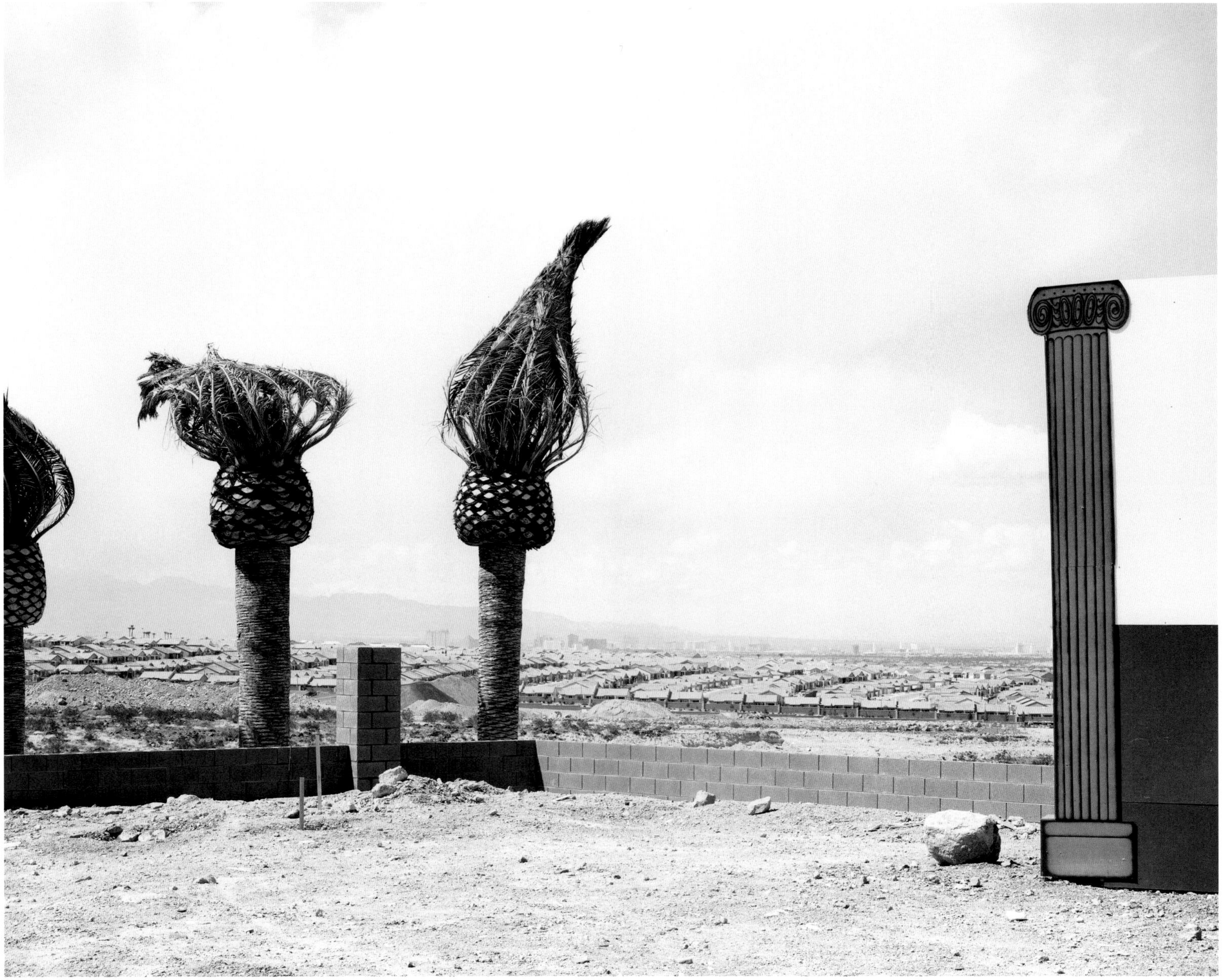

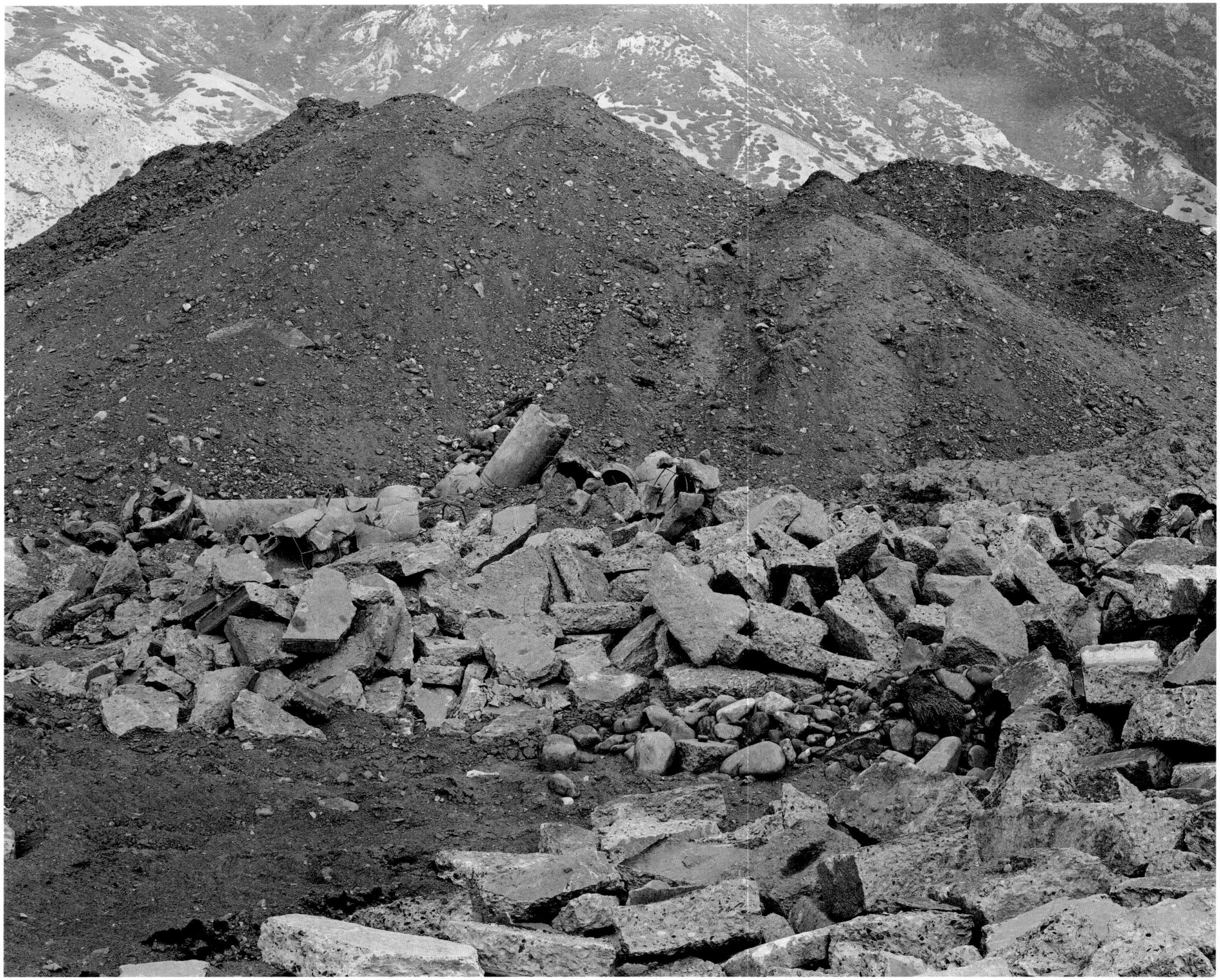

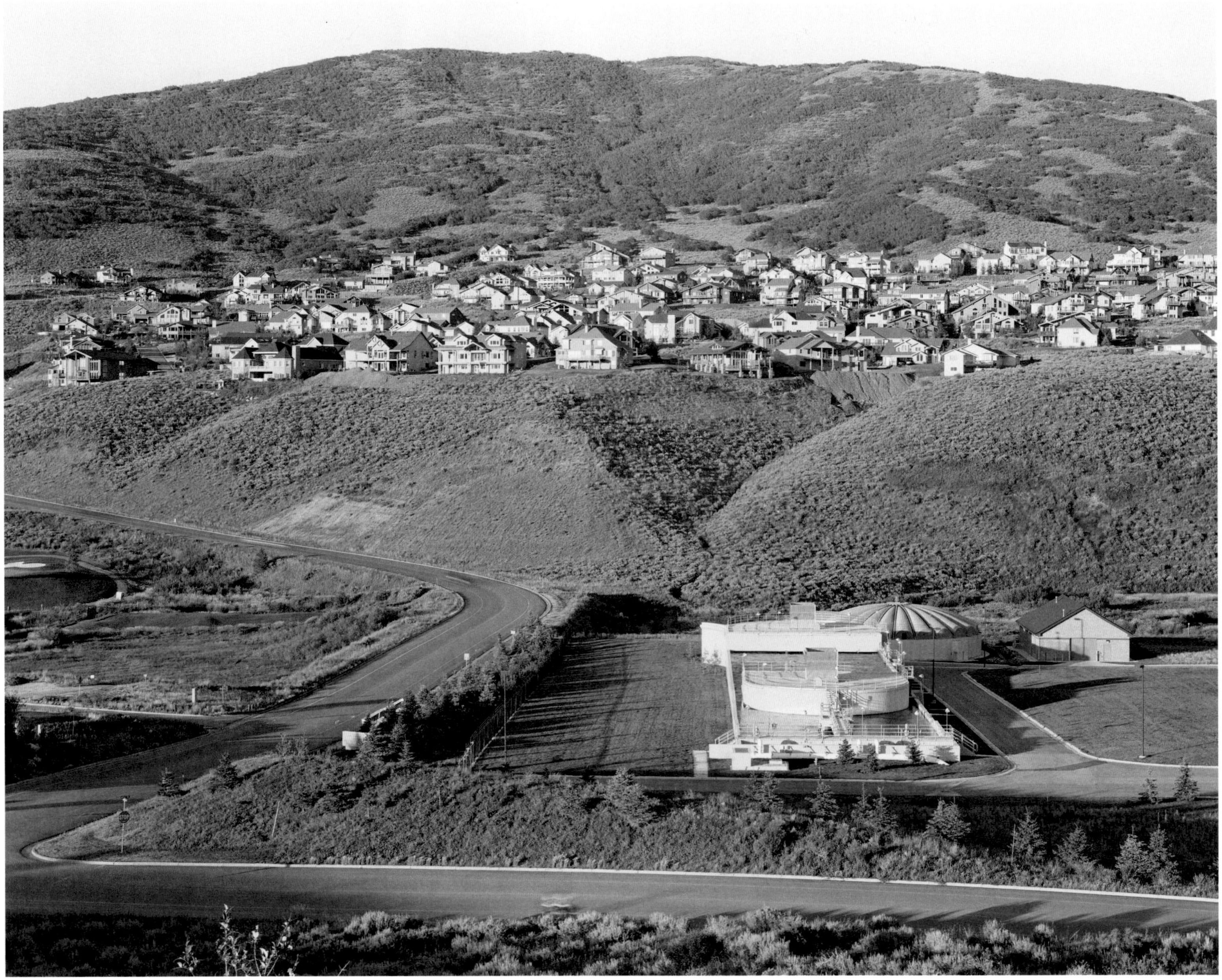

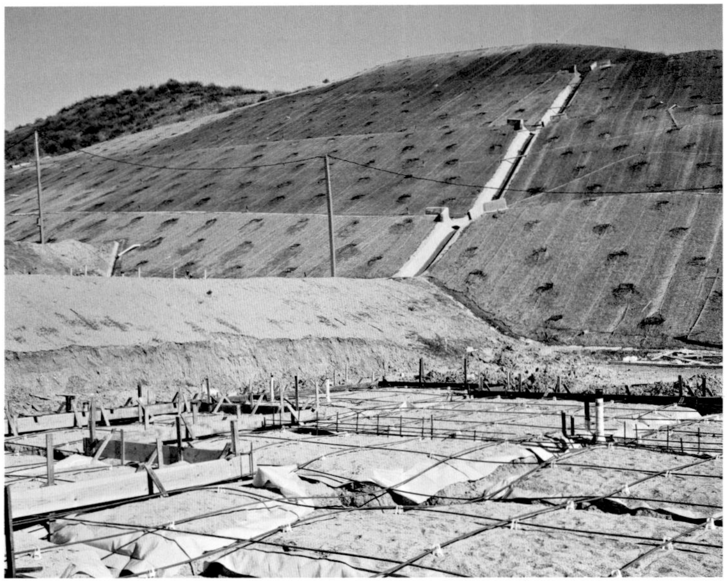

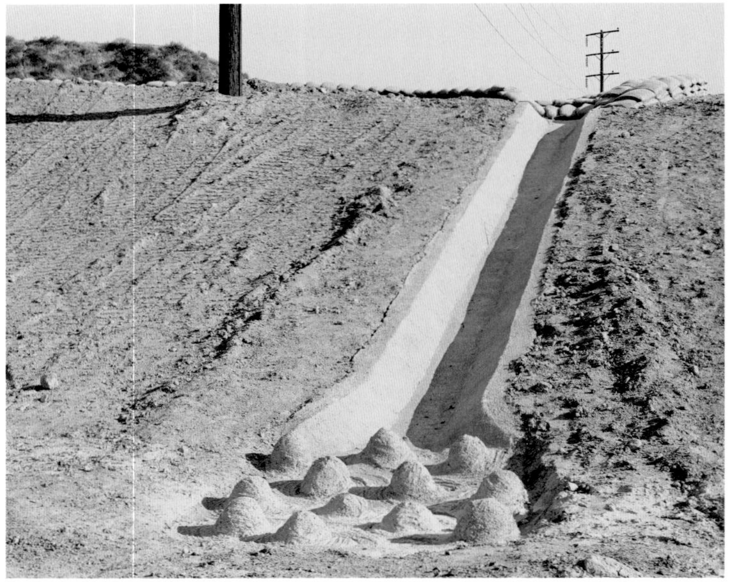

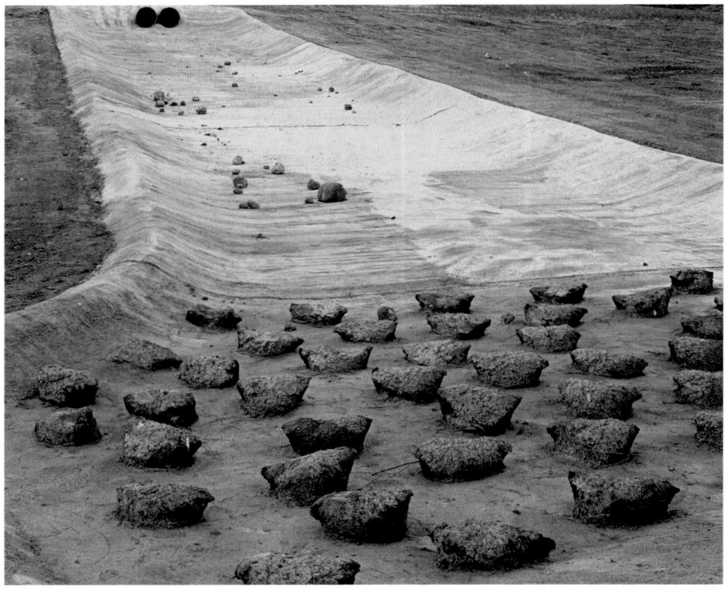

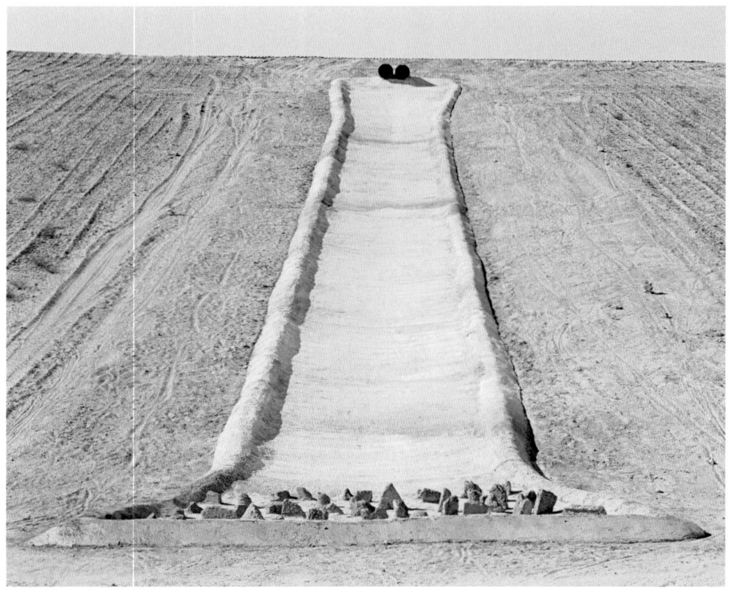

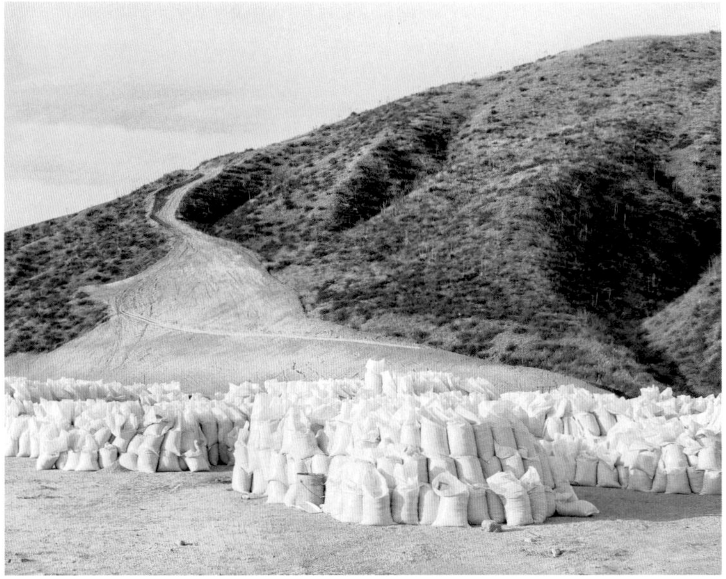
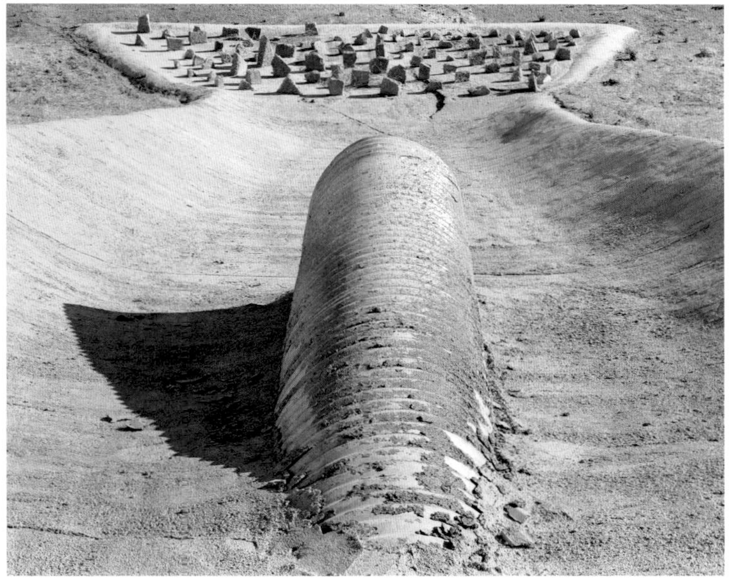
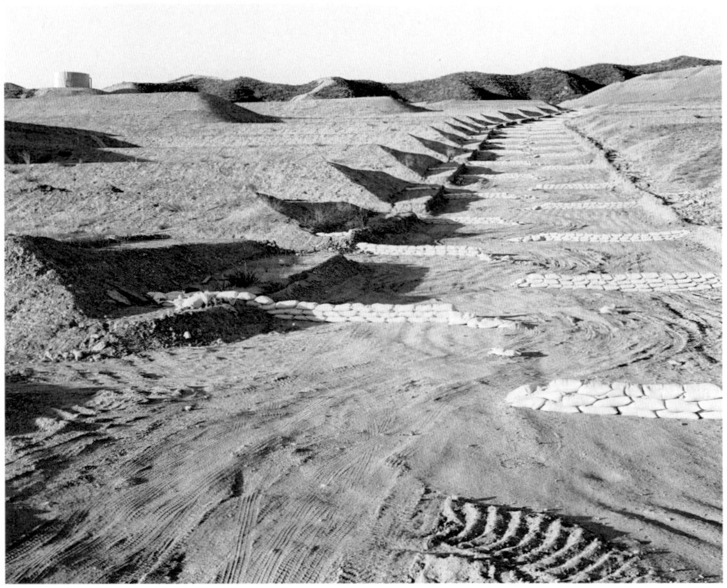
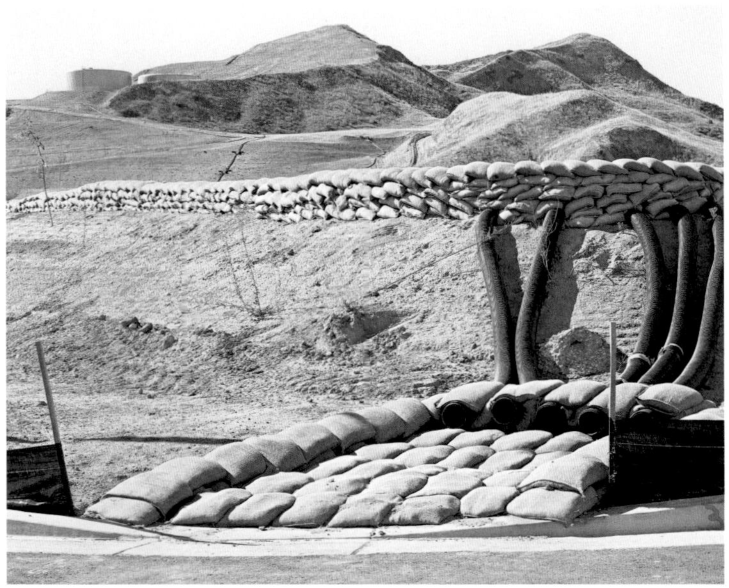

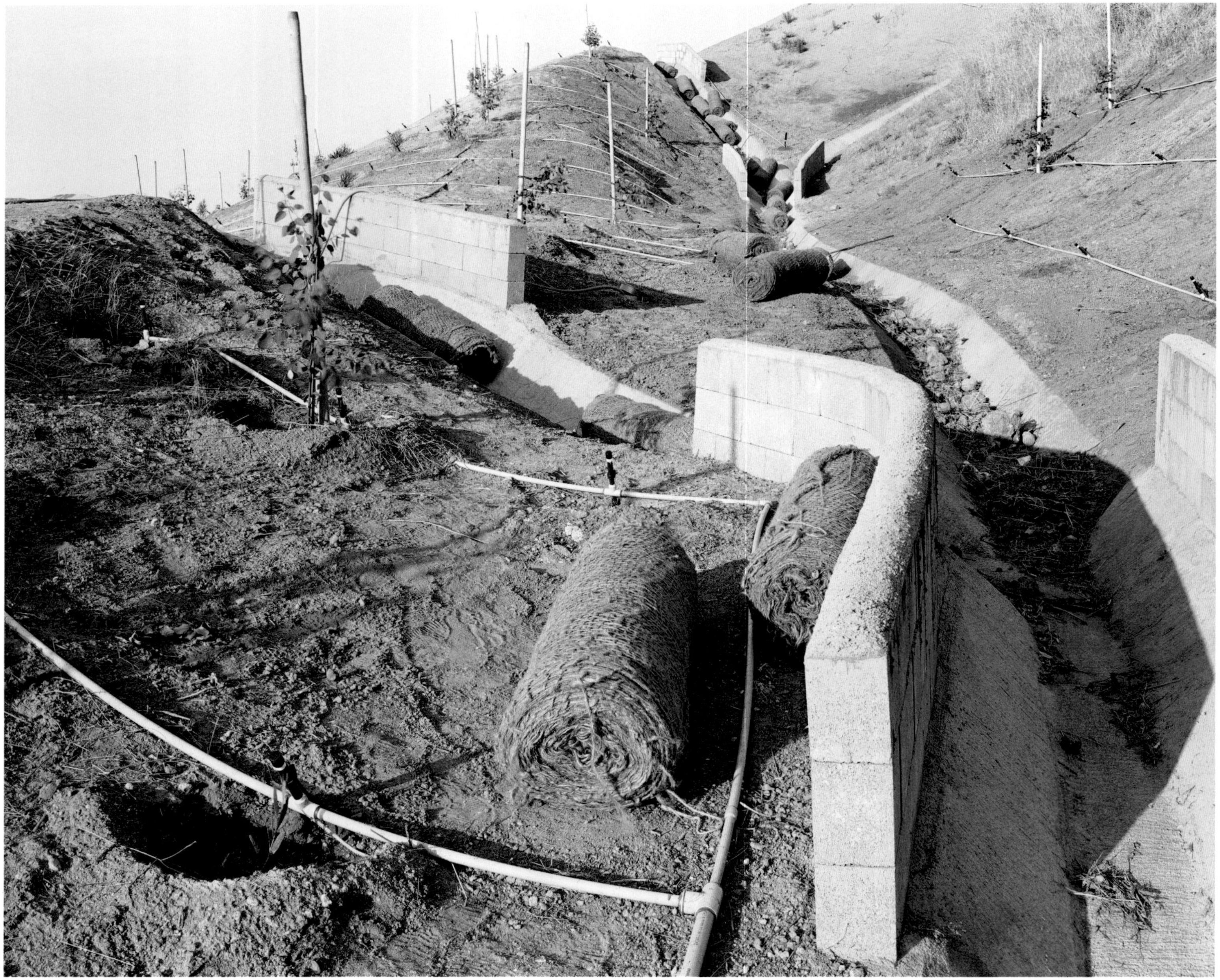

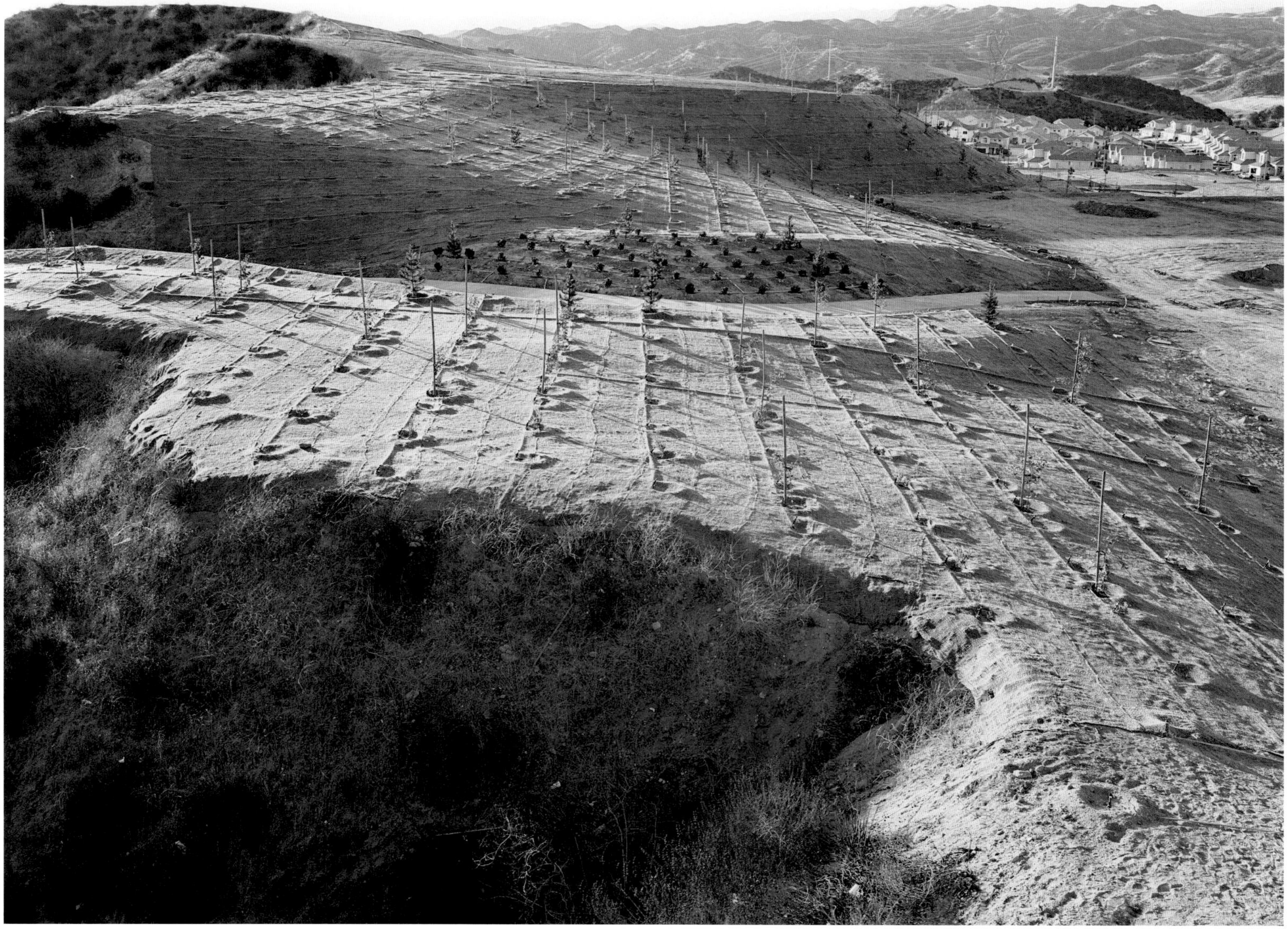

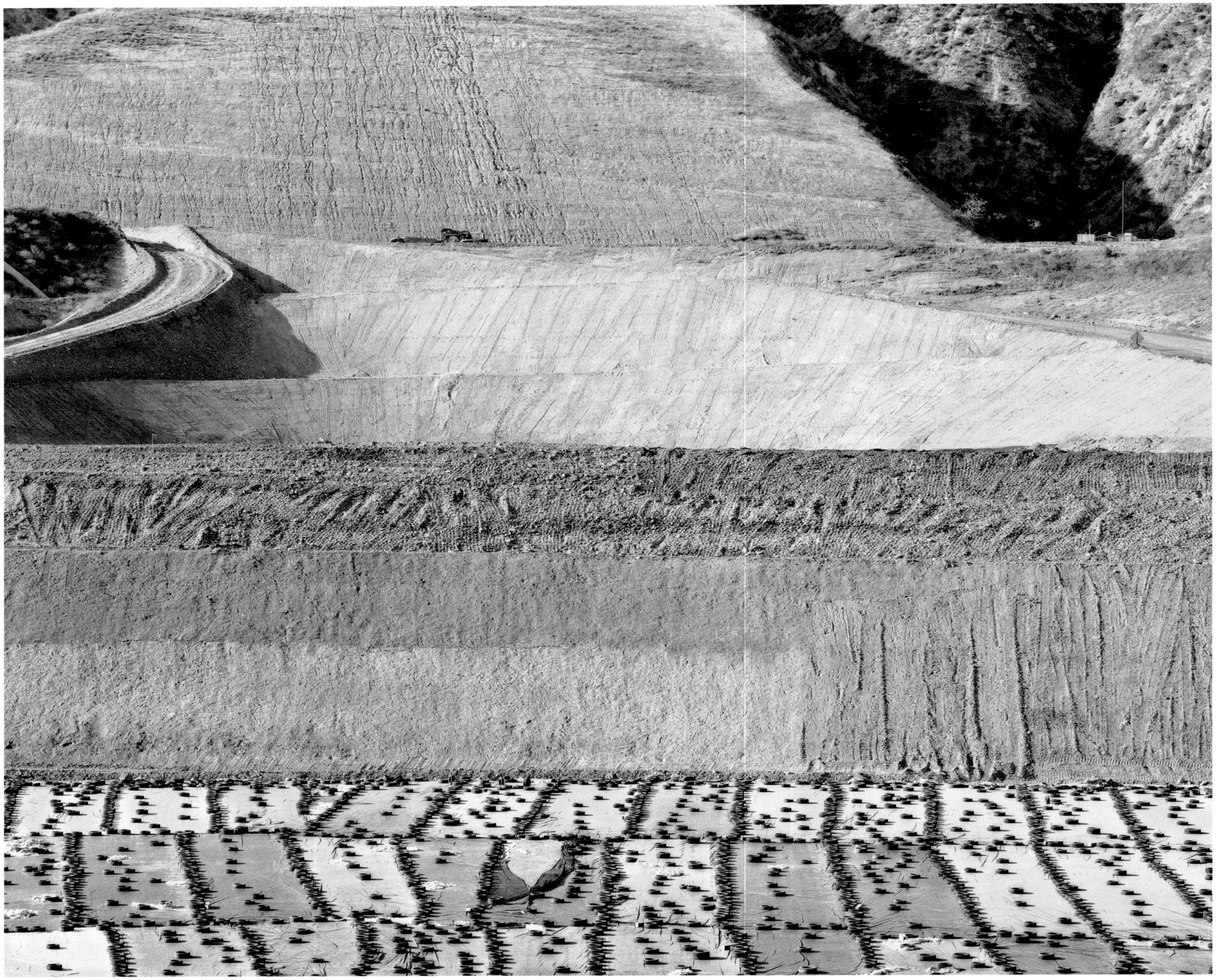

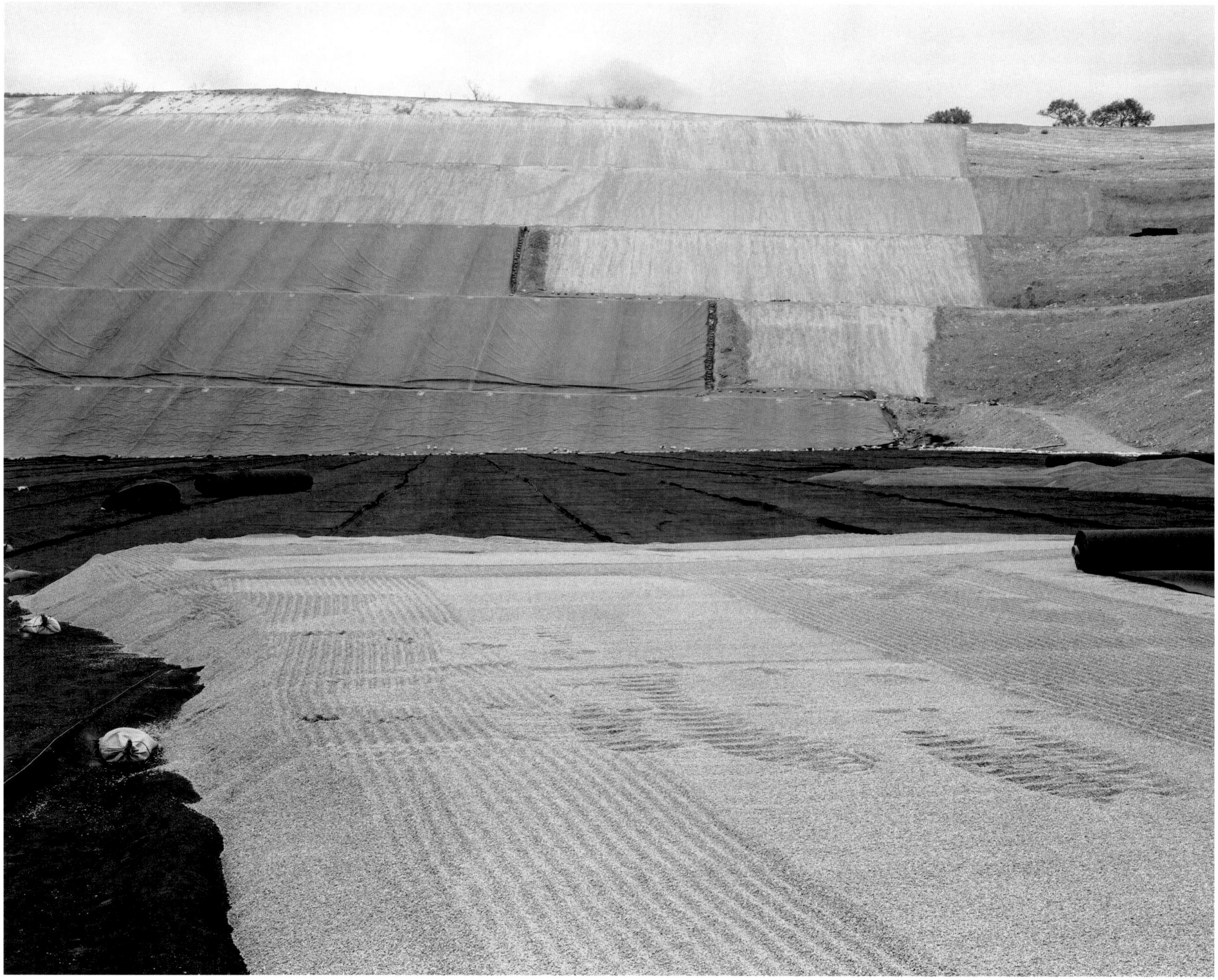

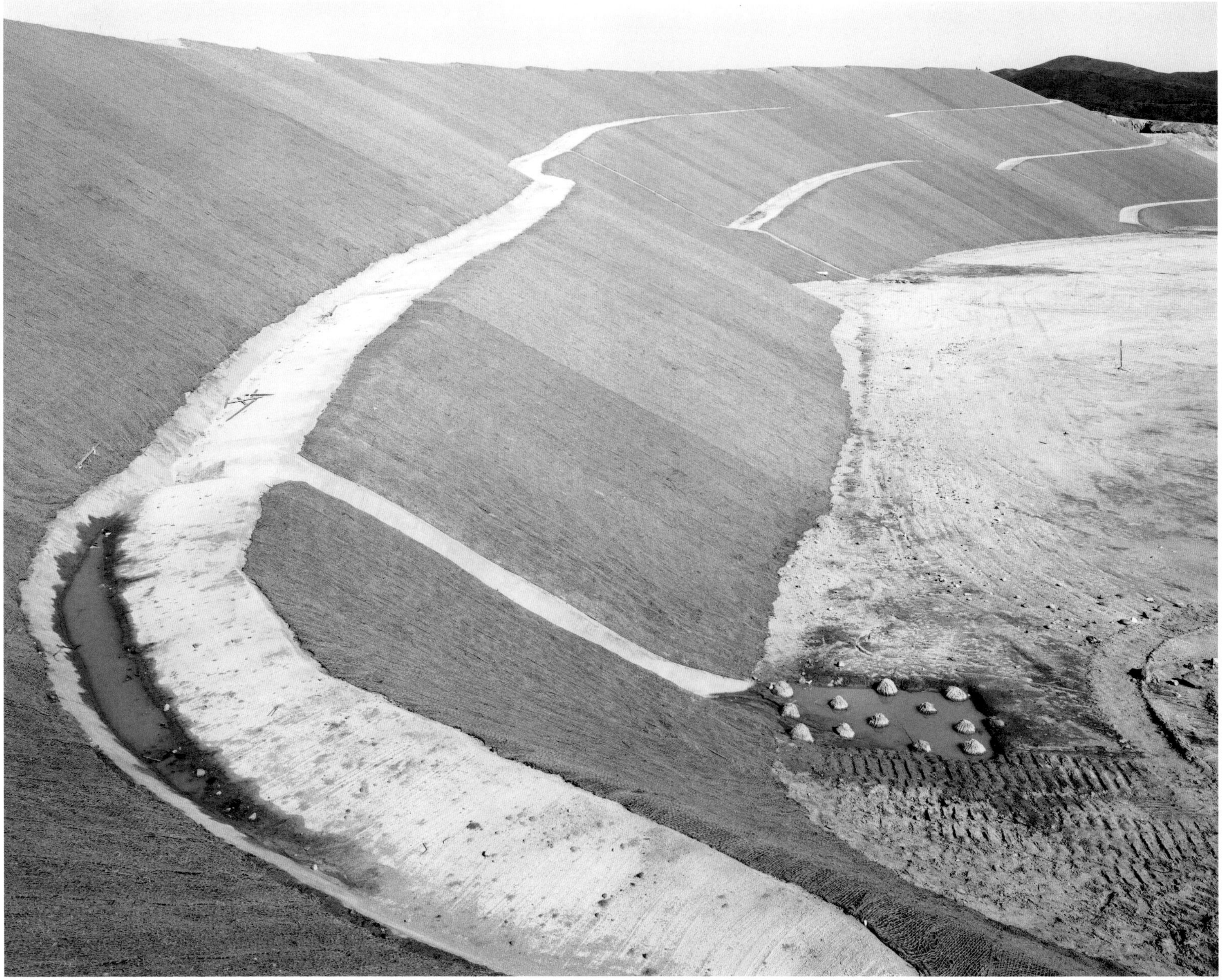

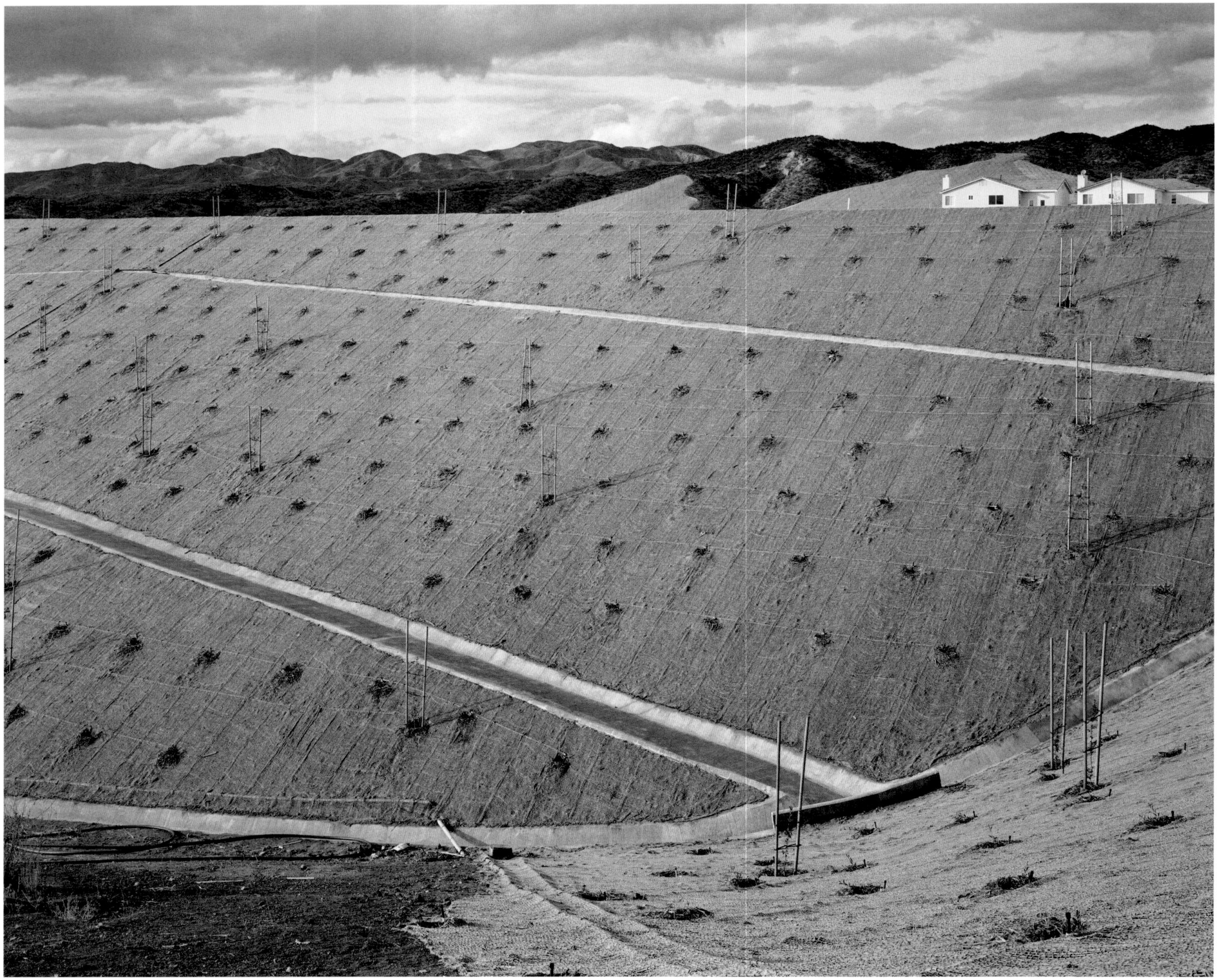

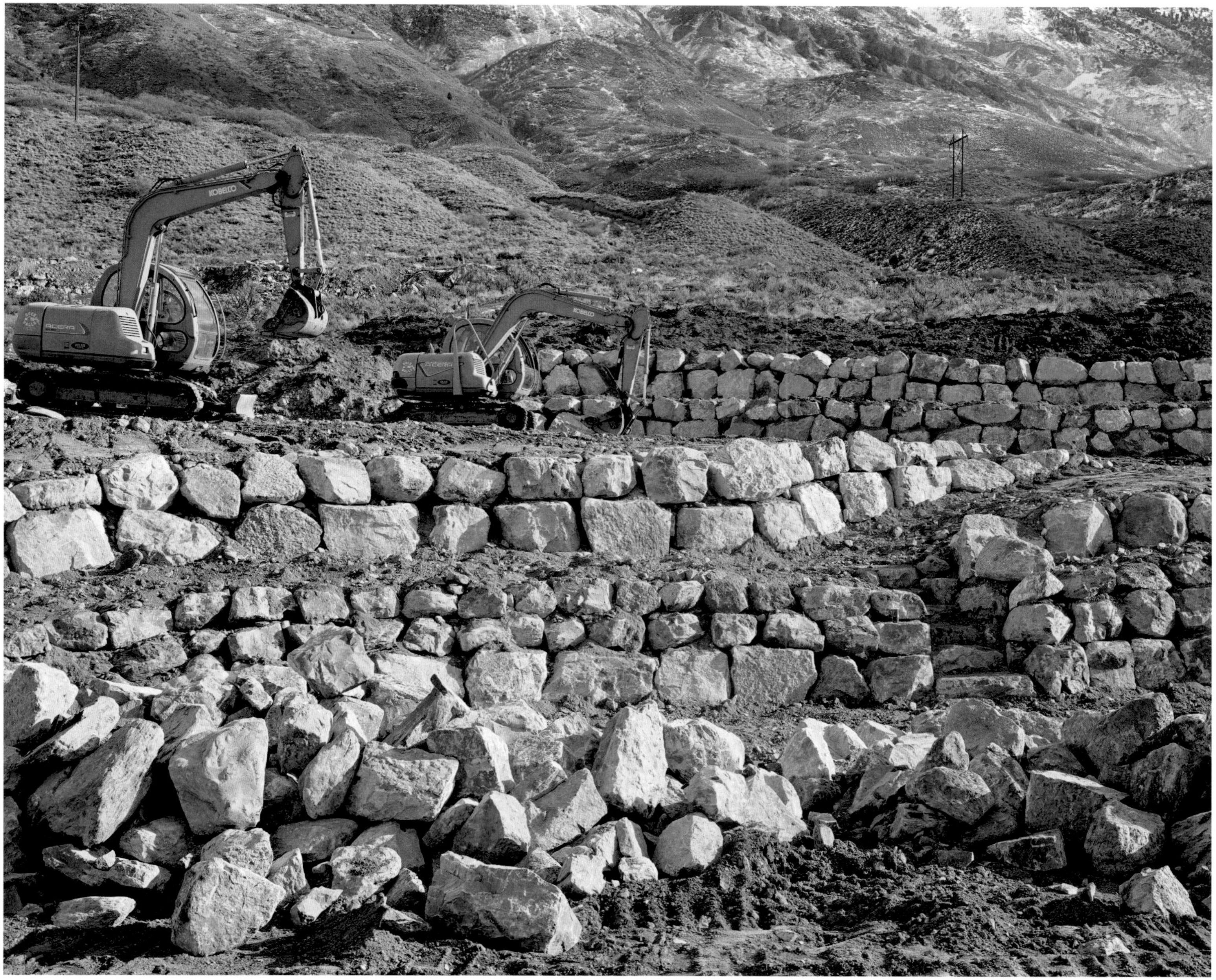

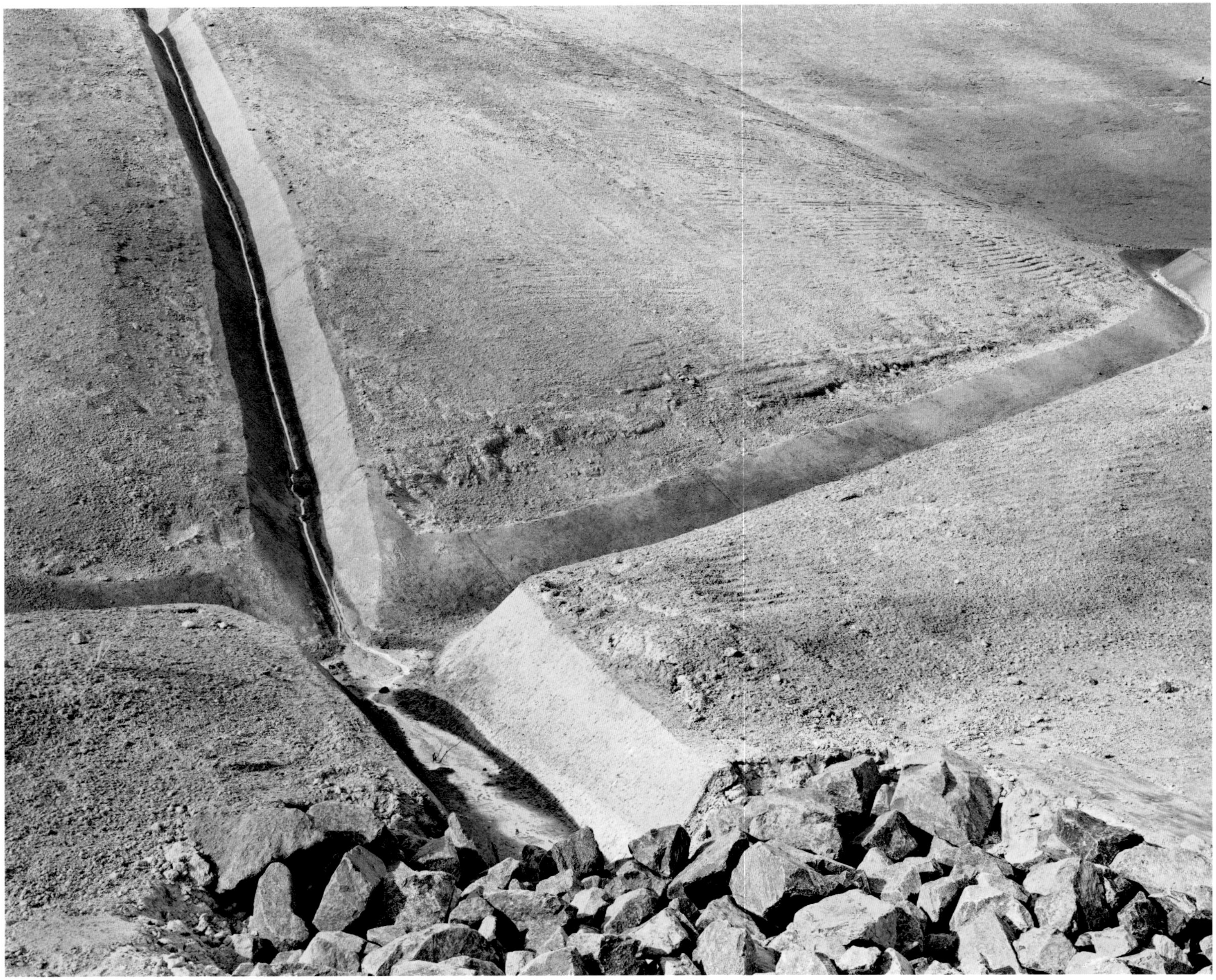

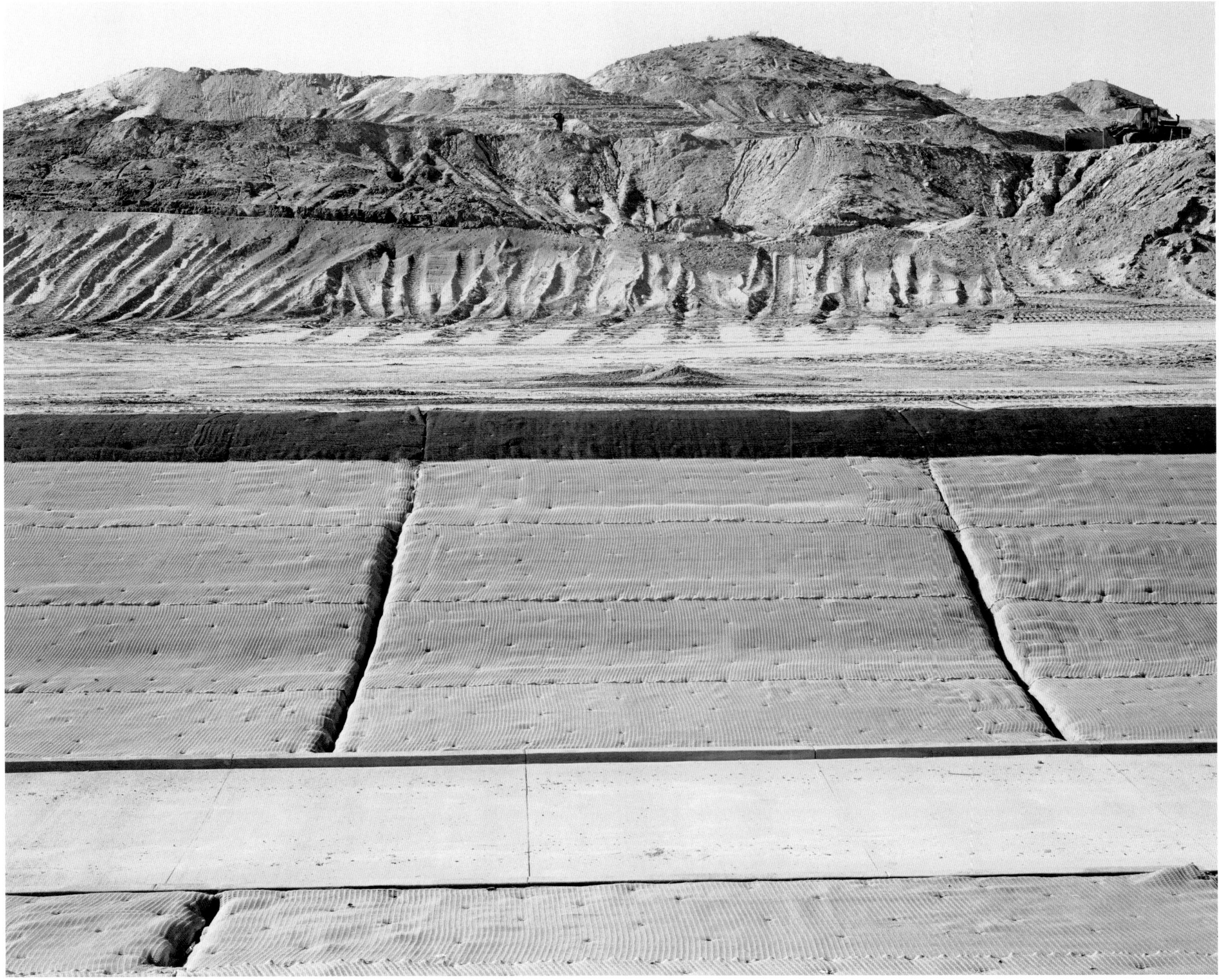

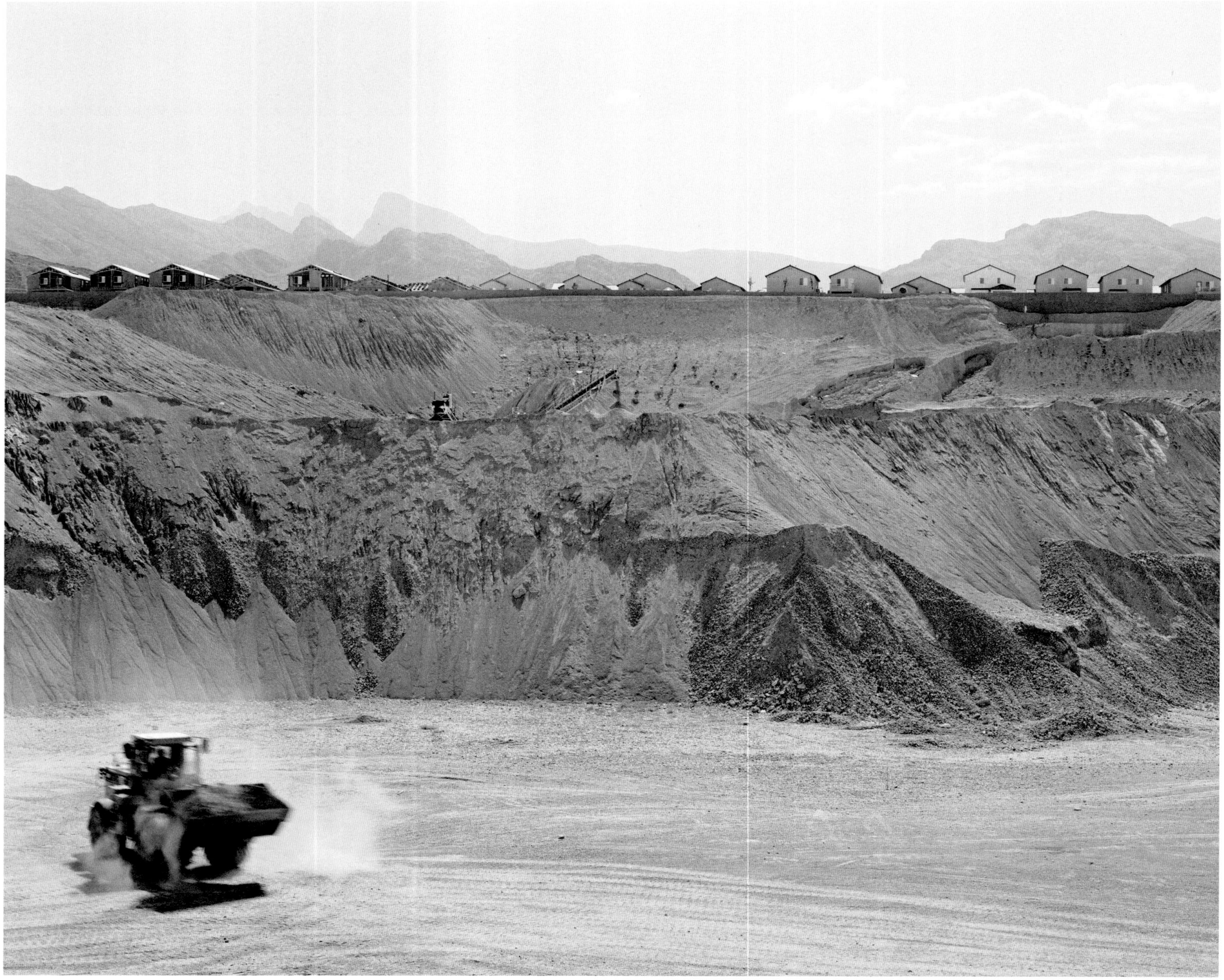

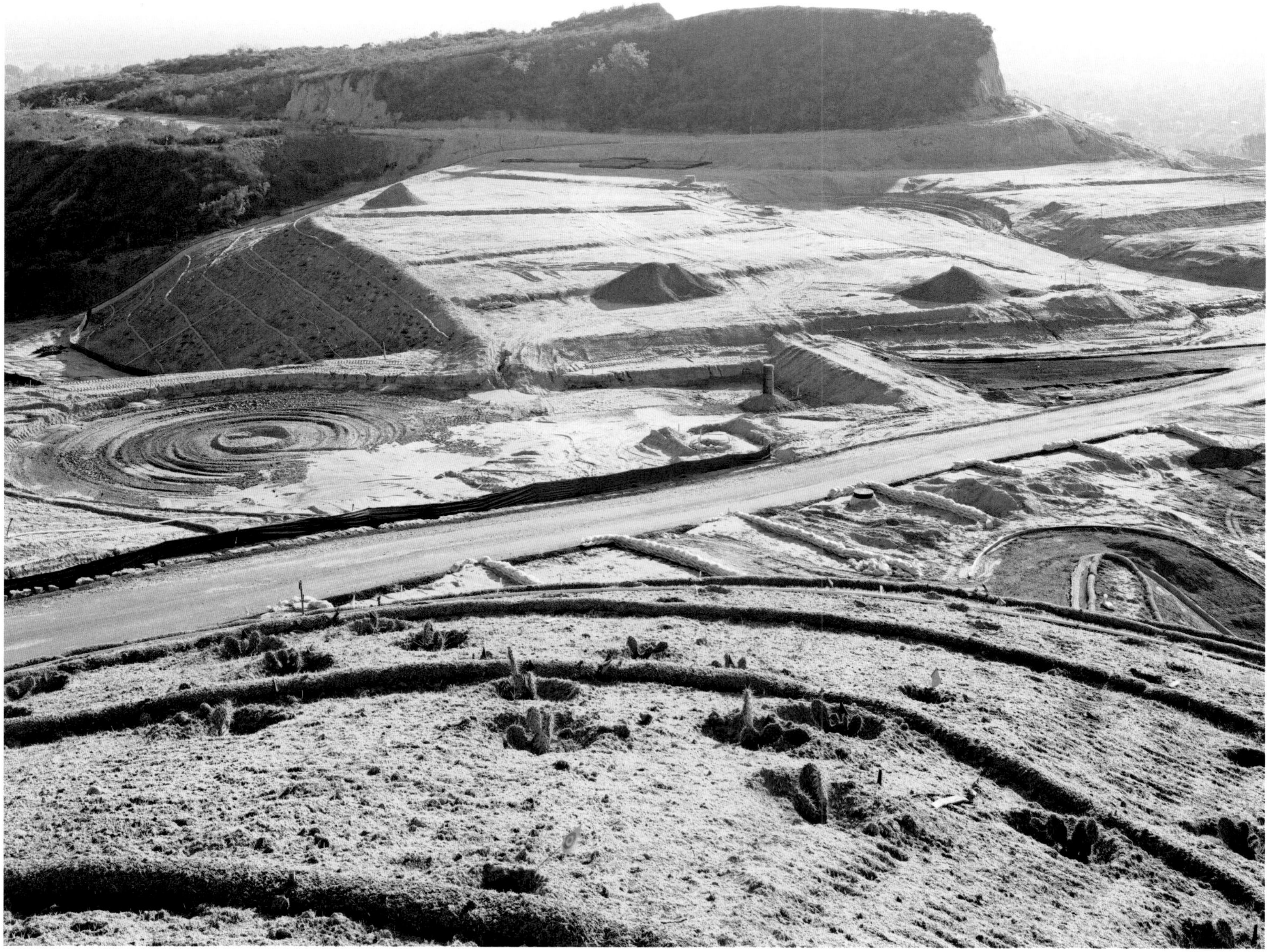

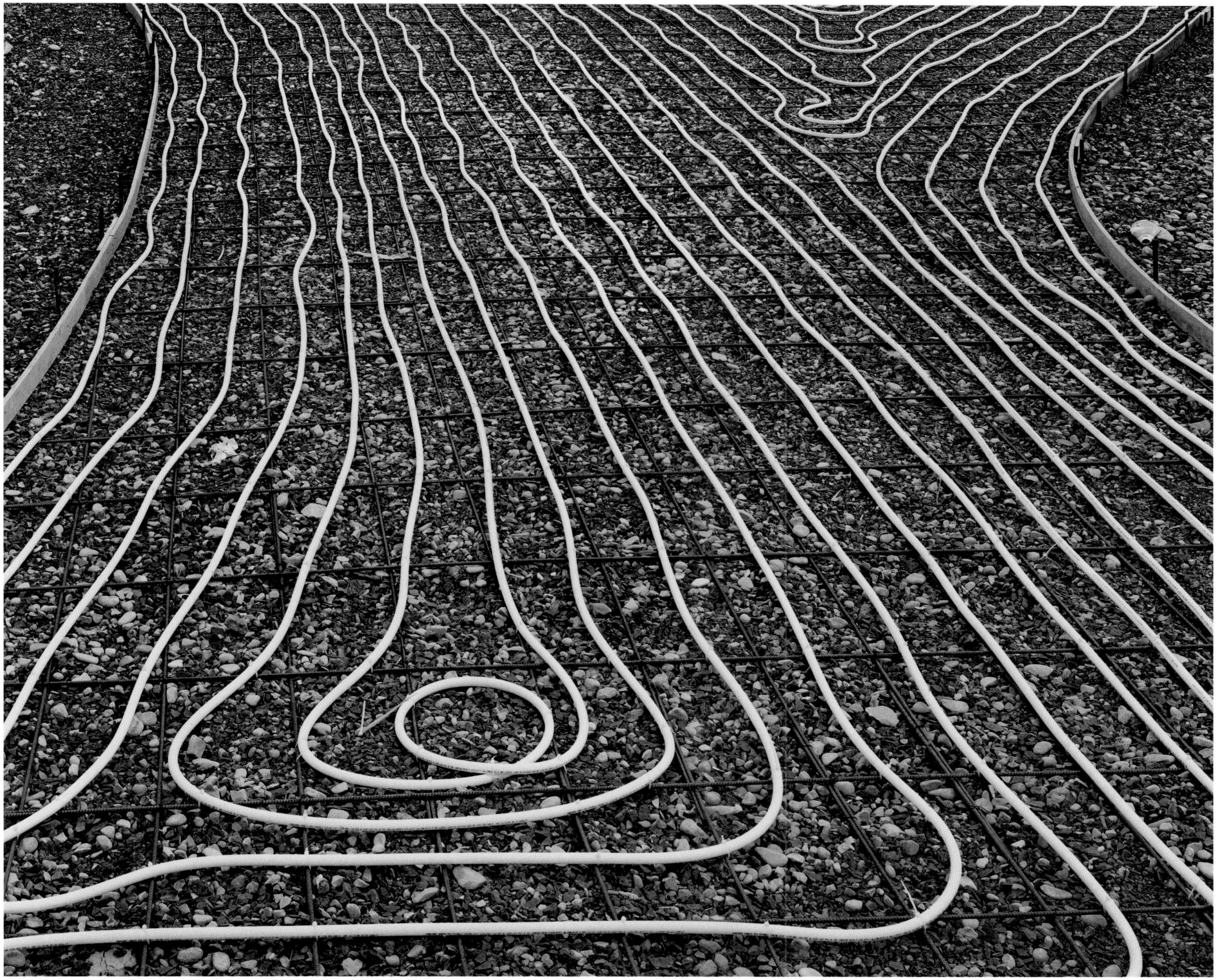

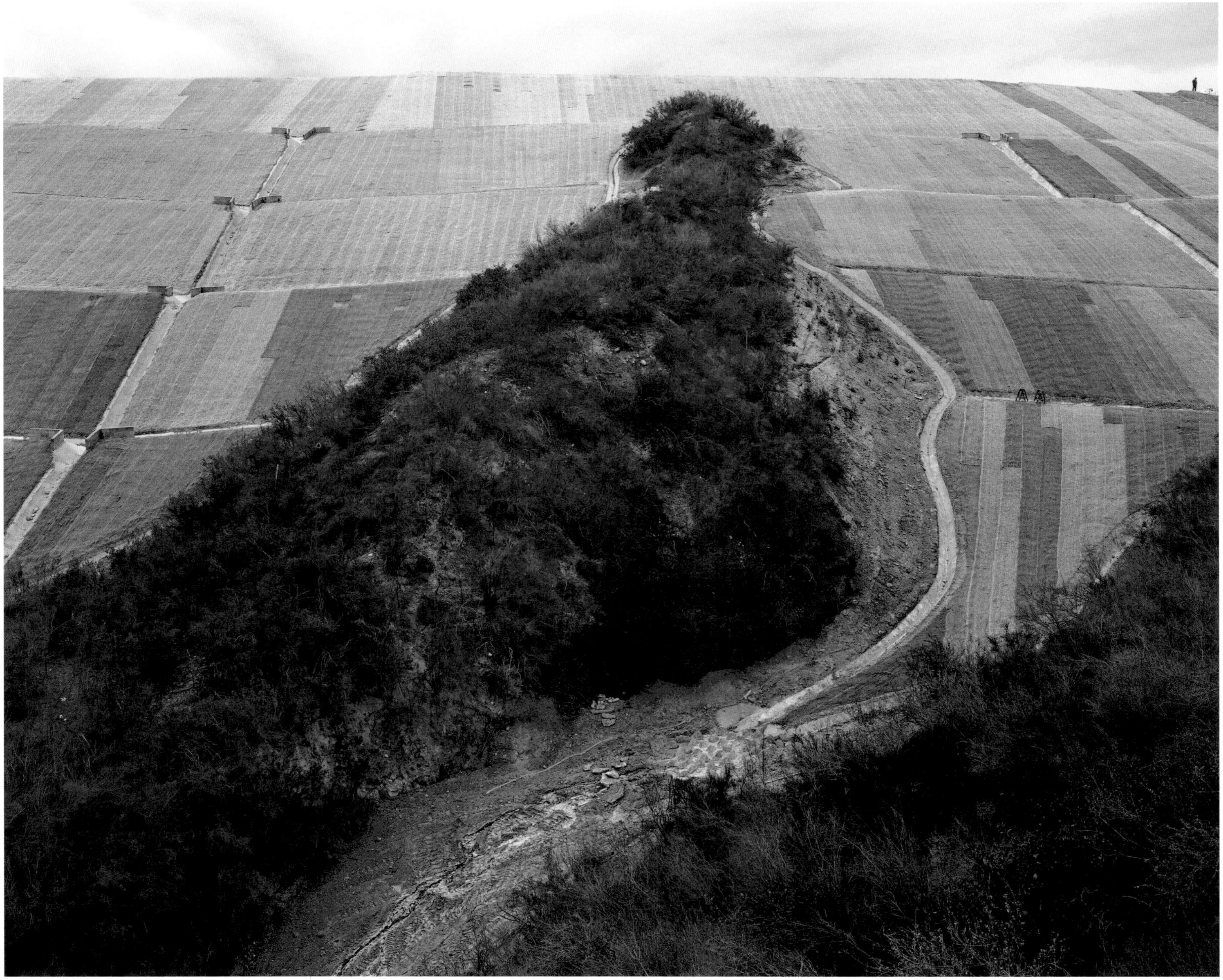

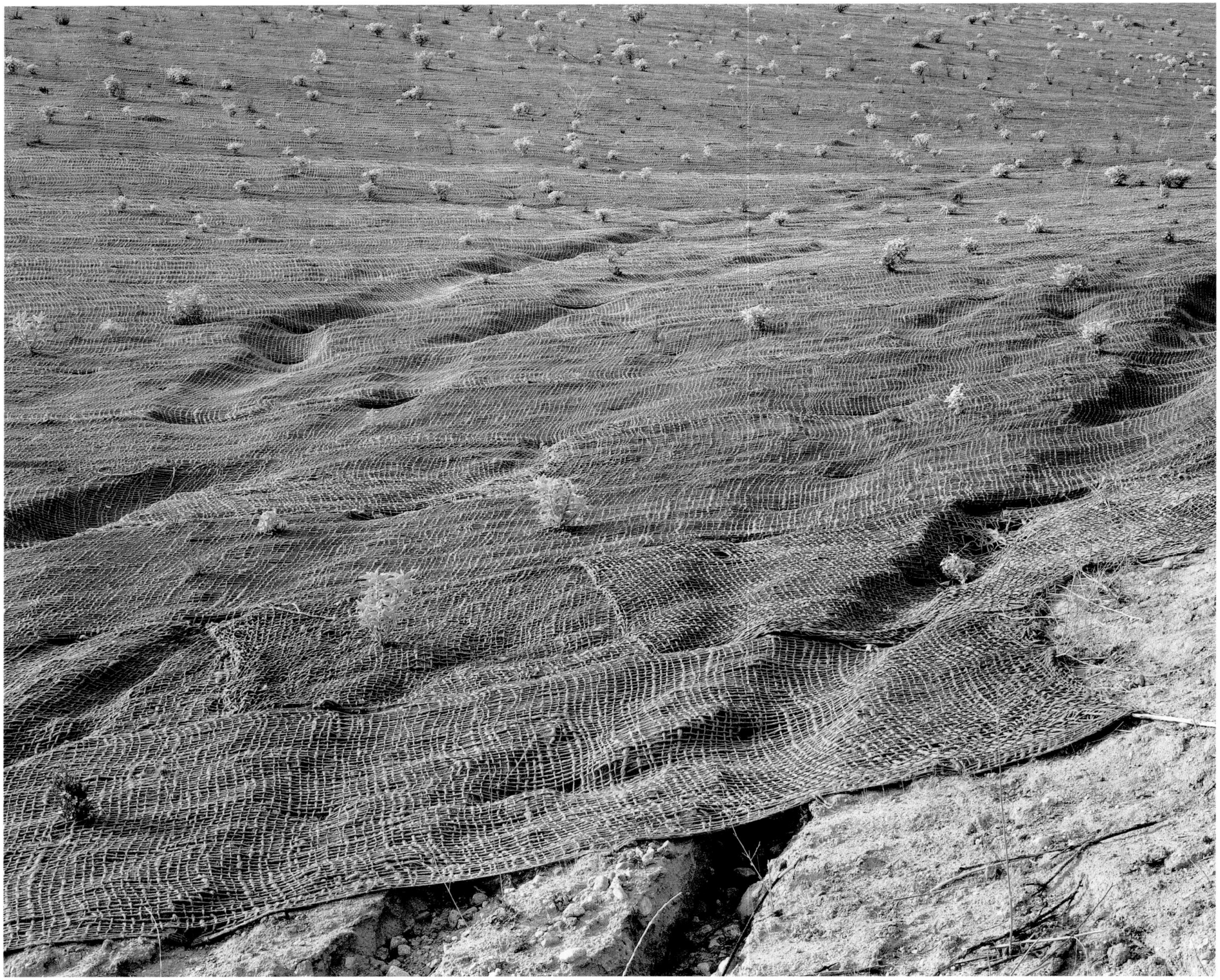

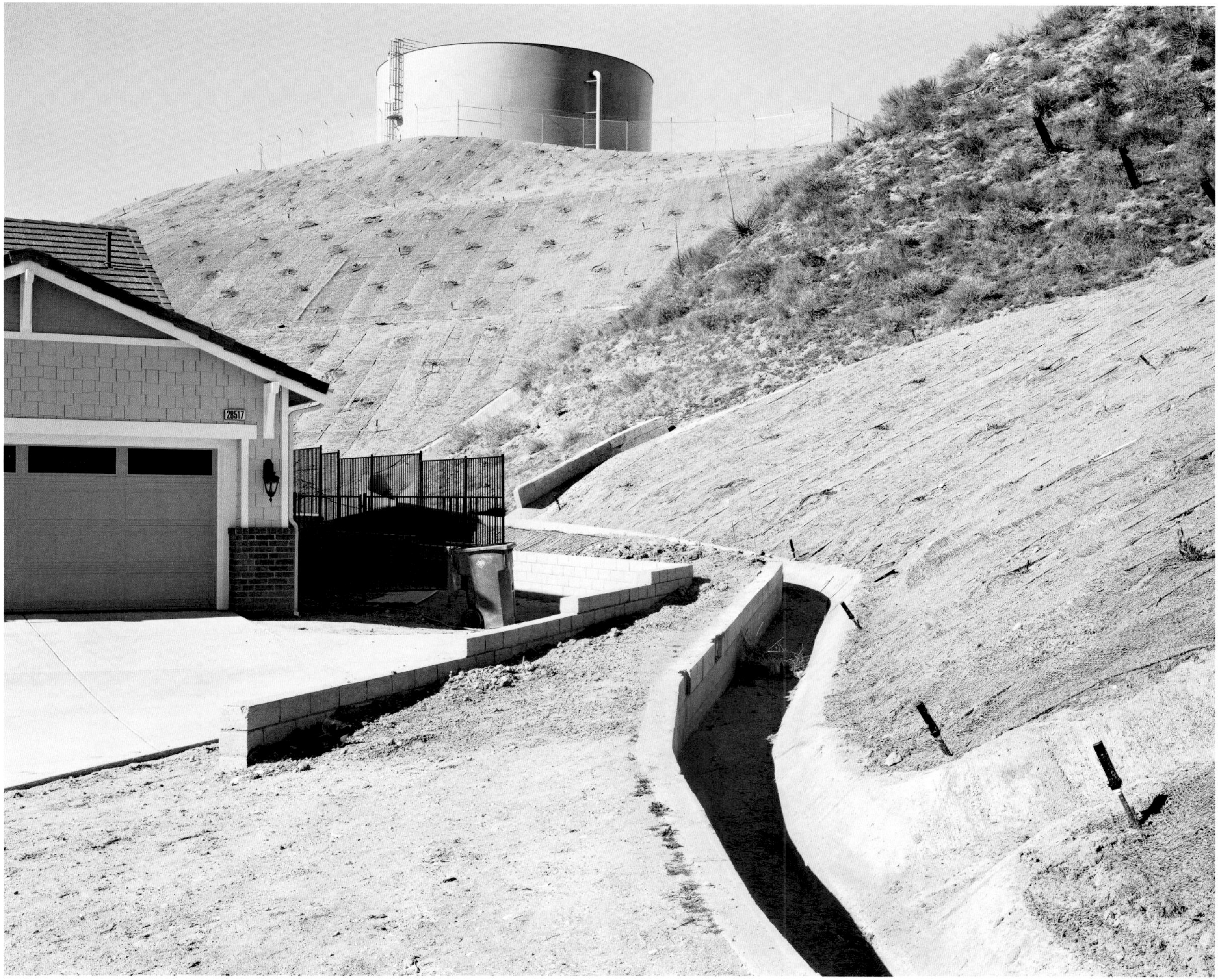
28517

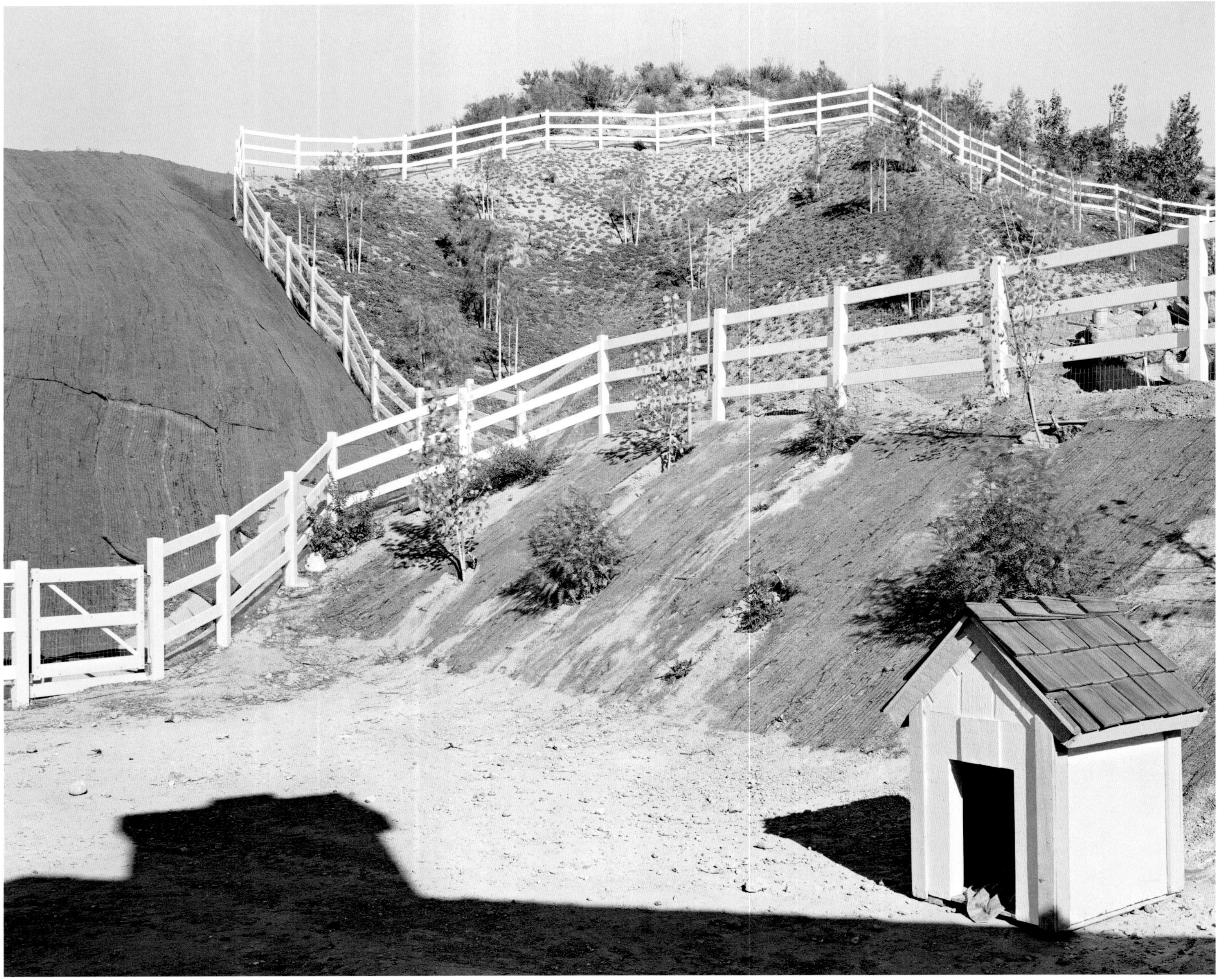

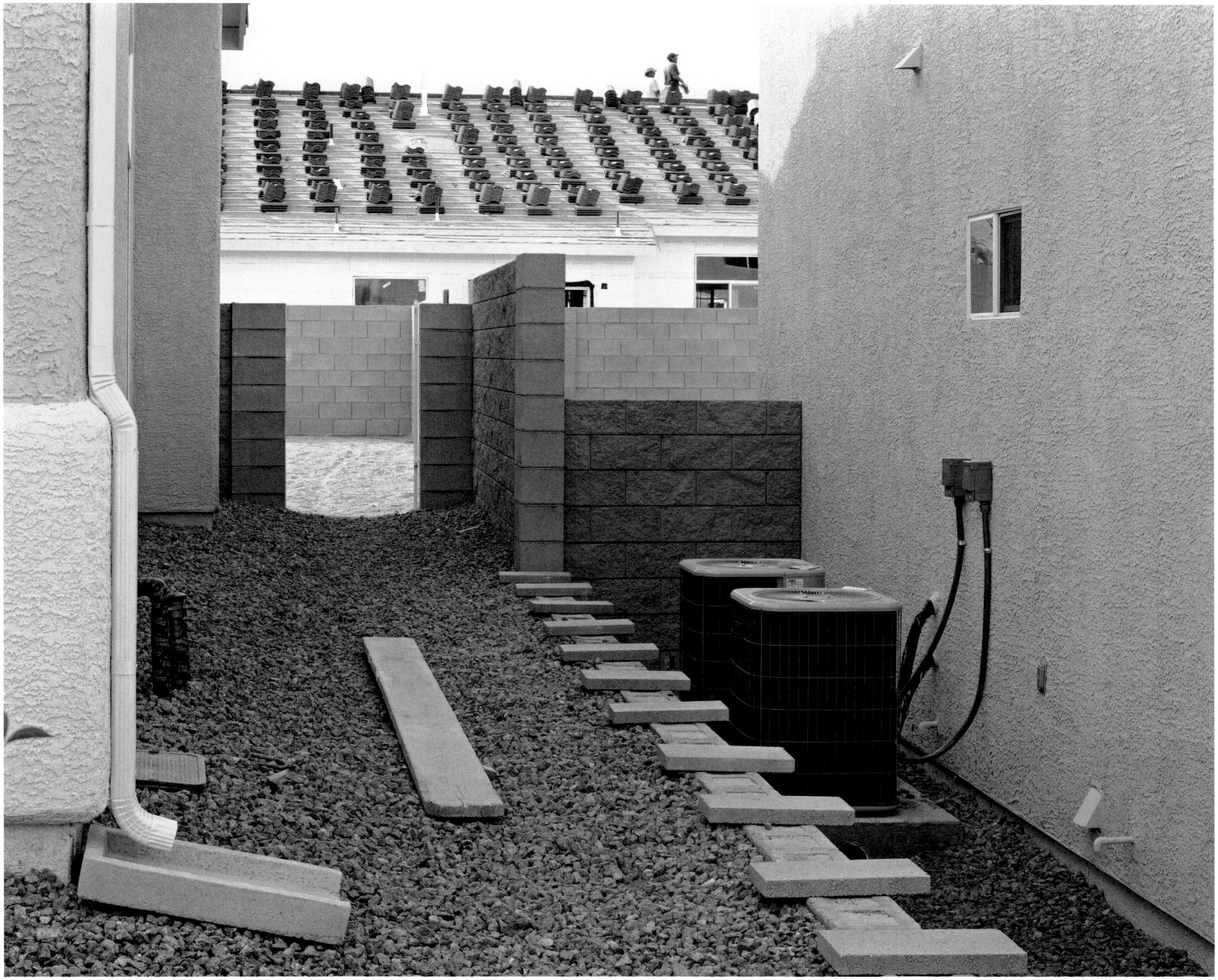

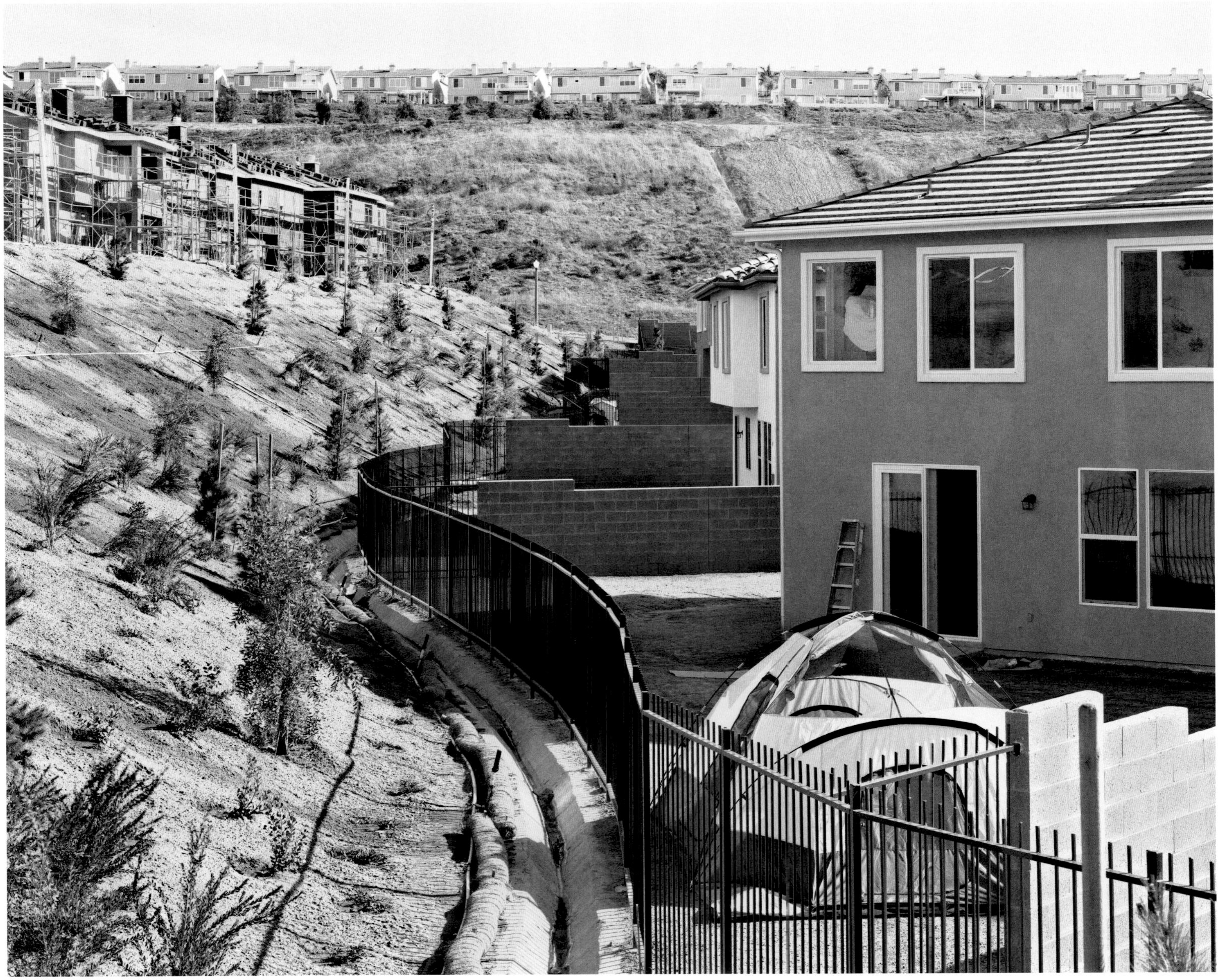

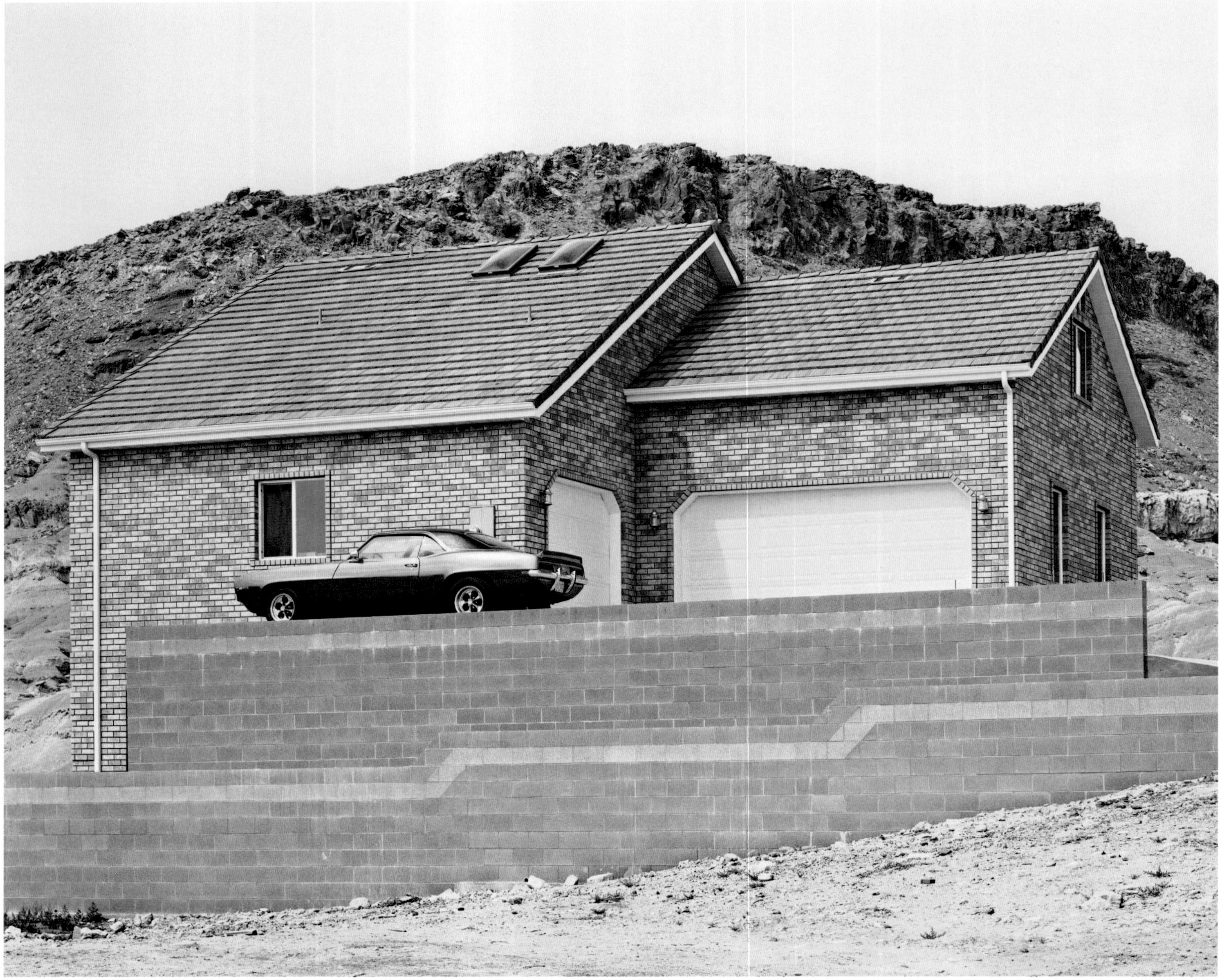

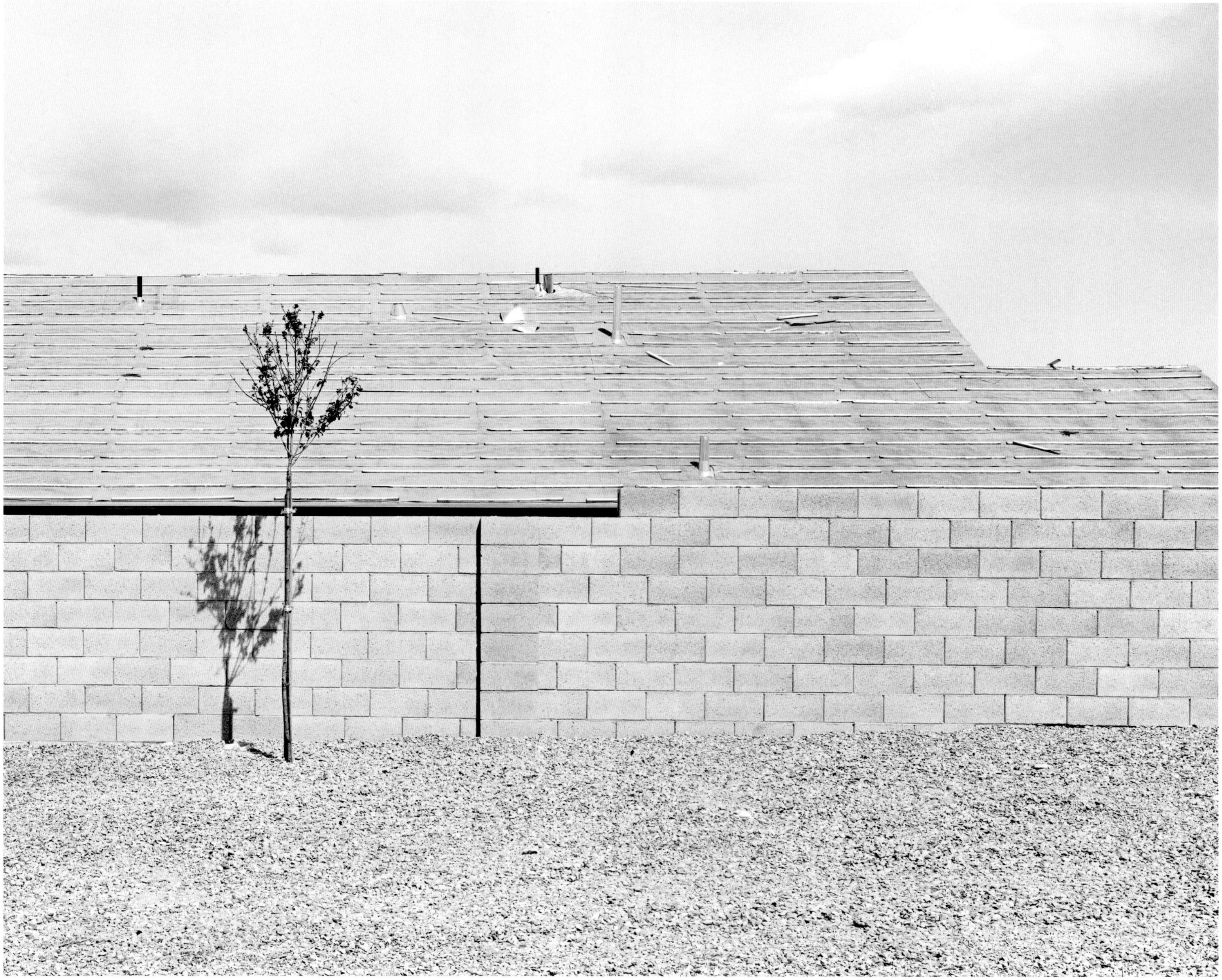

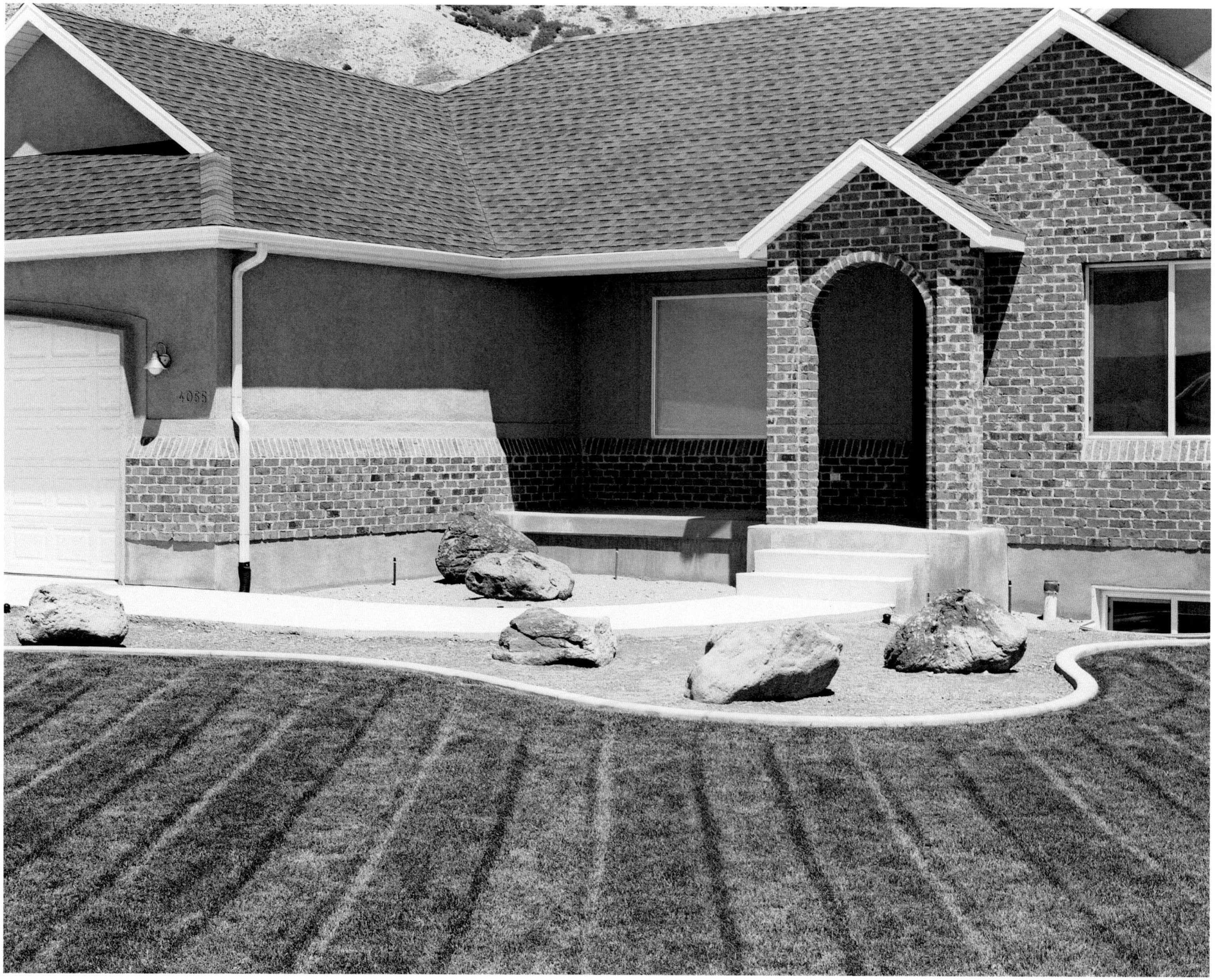

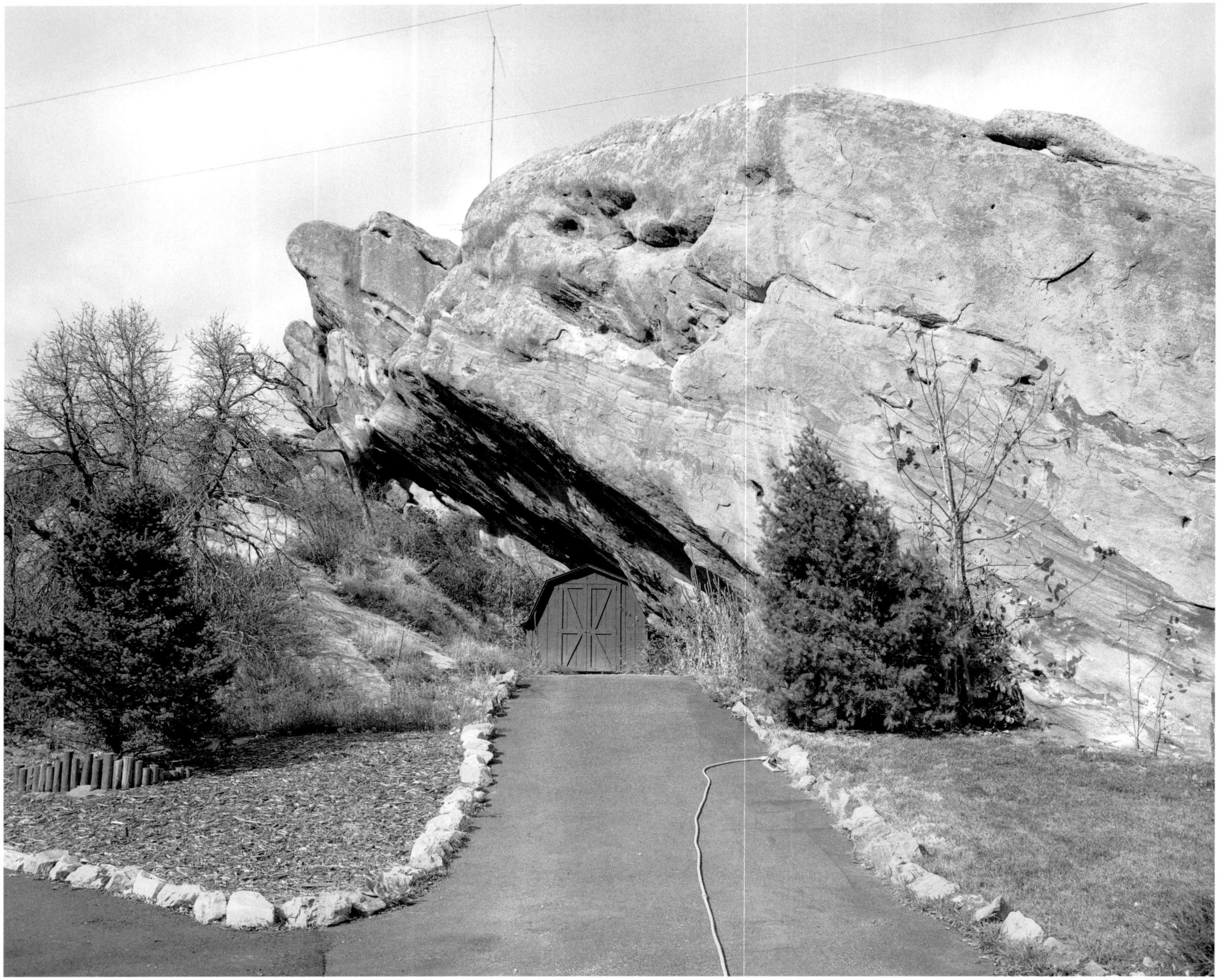

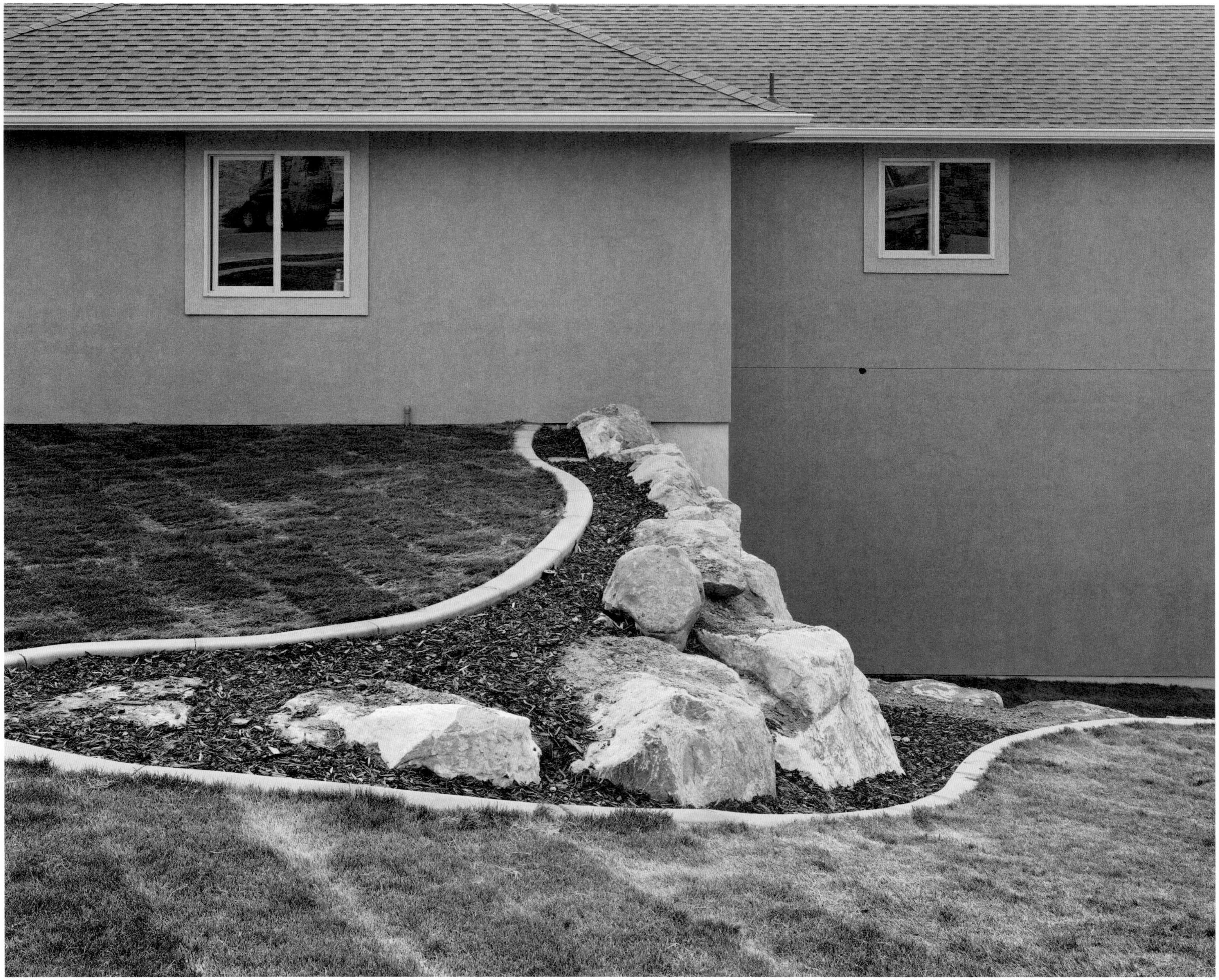

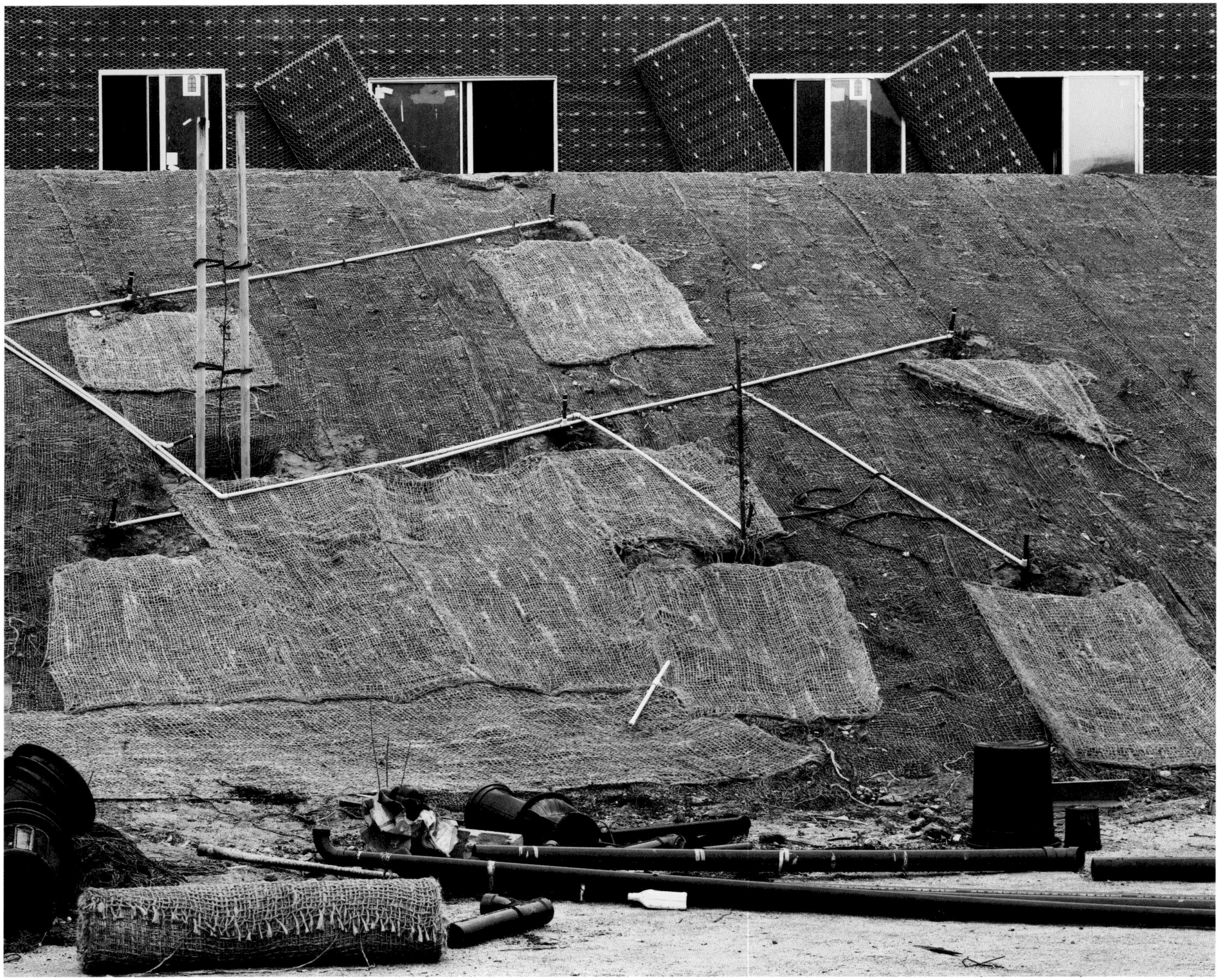

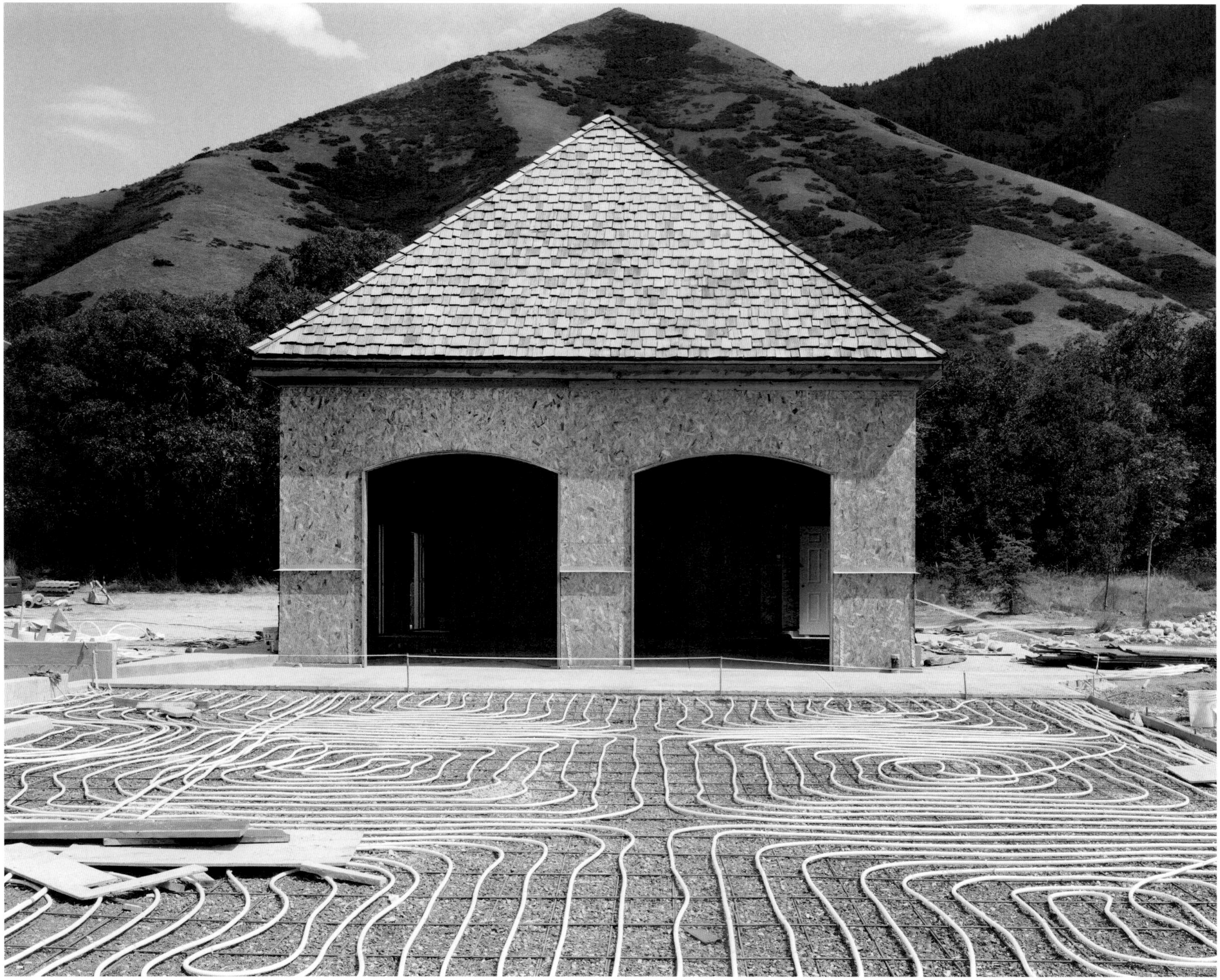

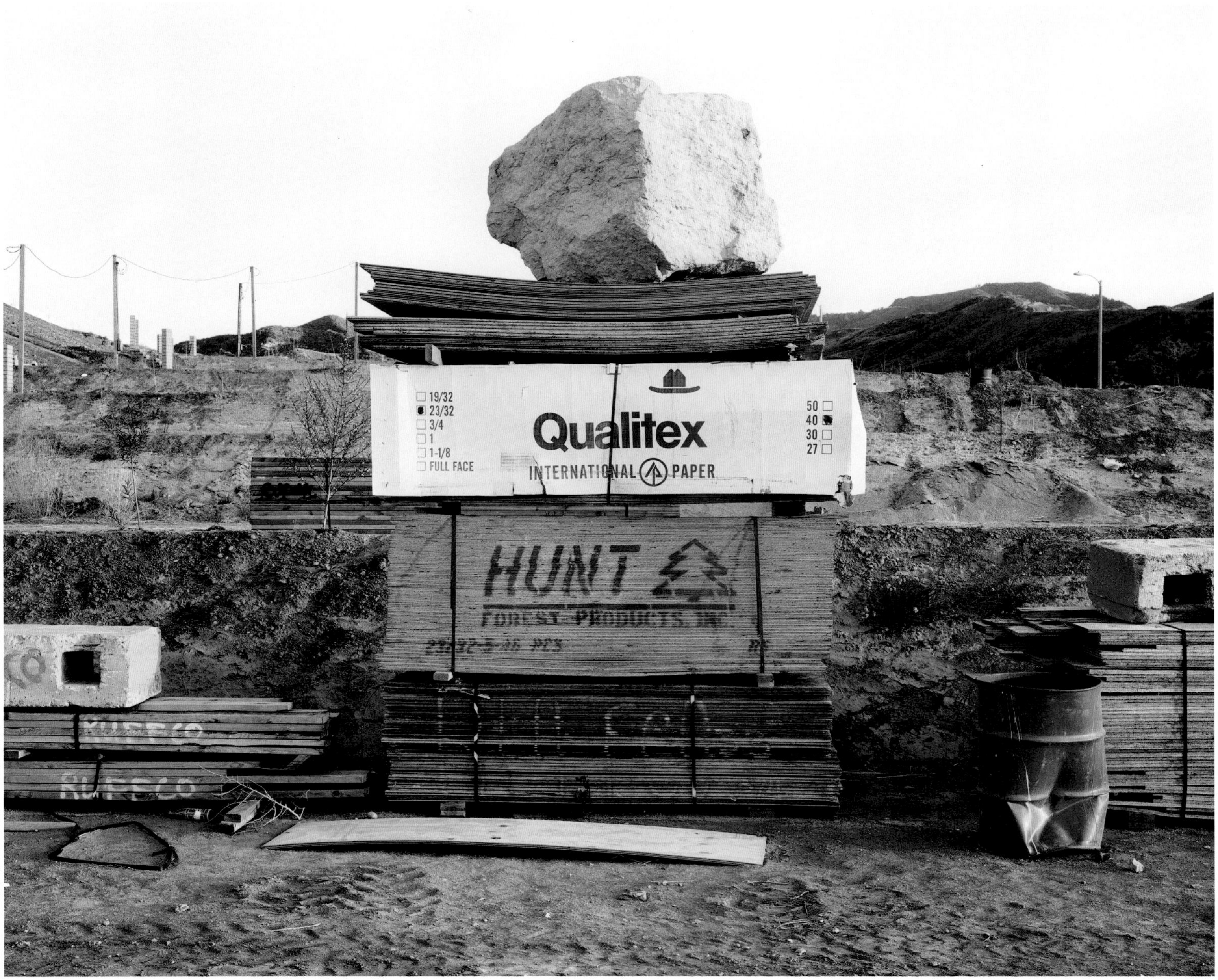
19/32
23/32
3/4
1
1-1/8
FULL FACE
Qualitex
INTERNATIONAL PAPER
50
40
30
27
HUNT
FOREST PRODUCTS, INC.

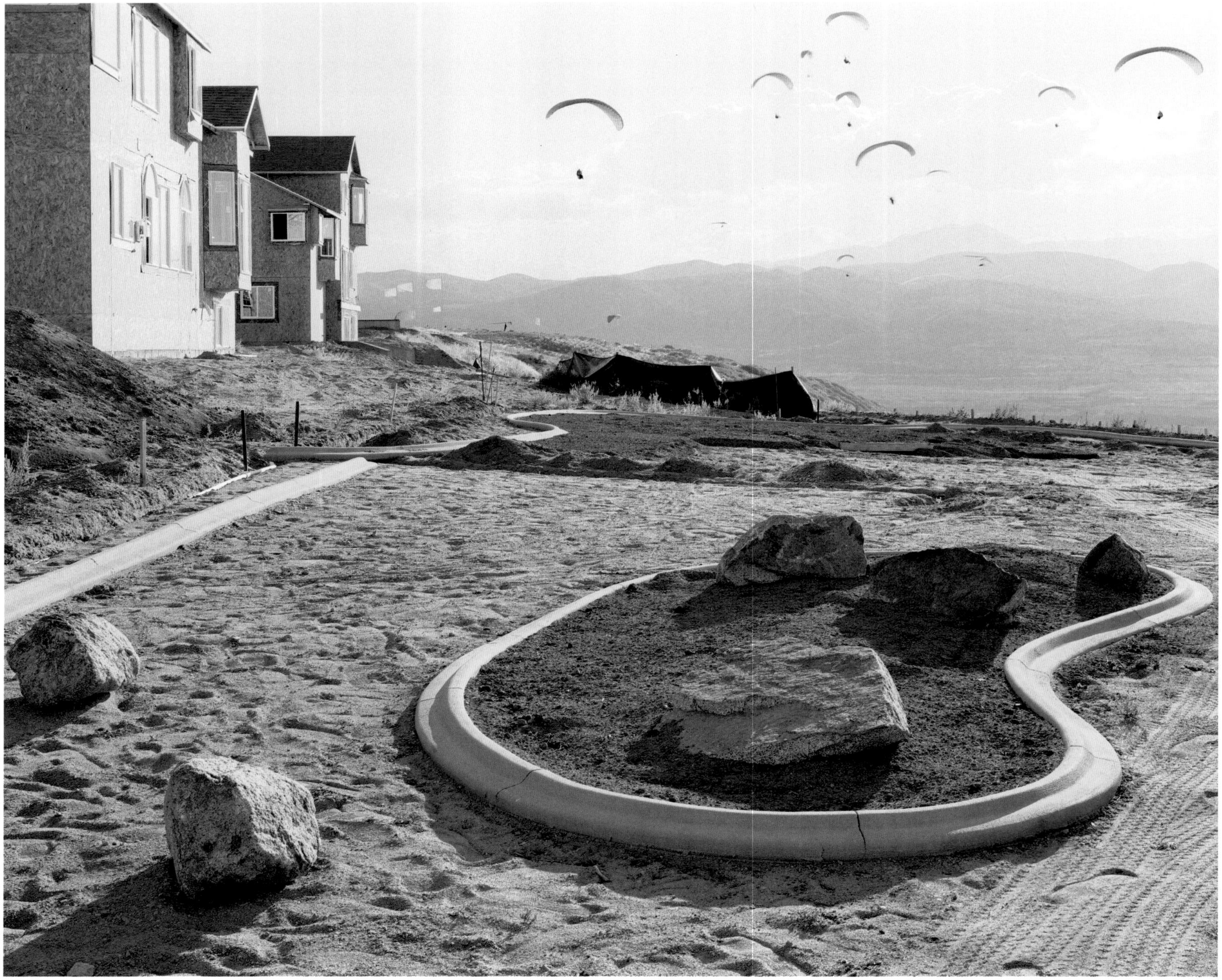

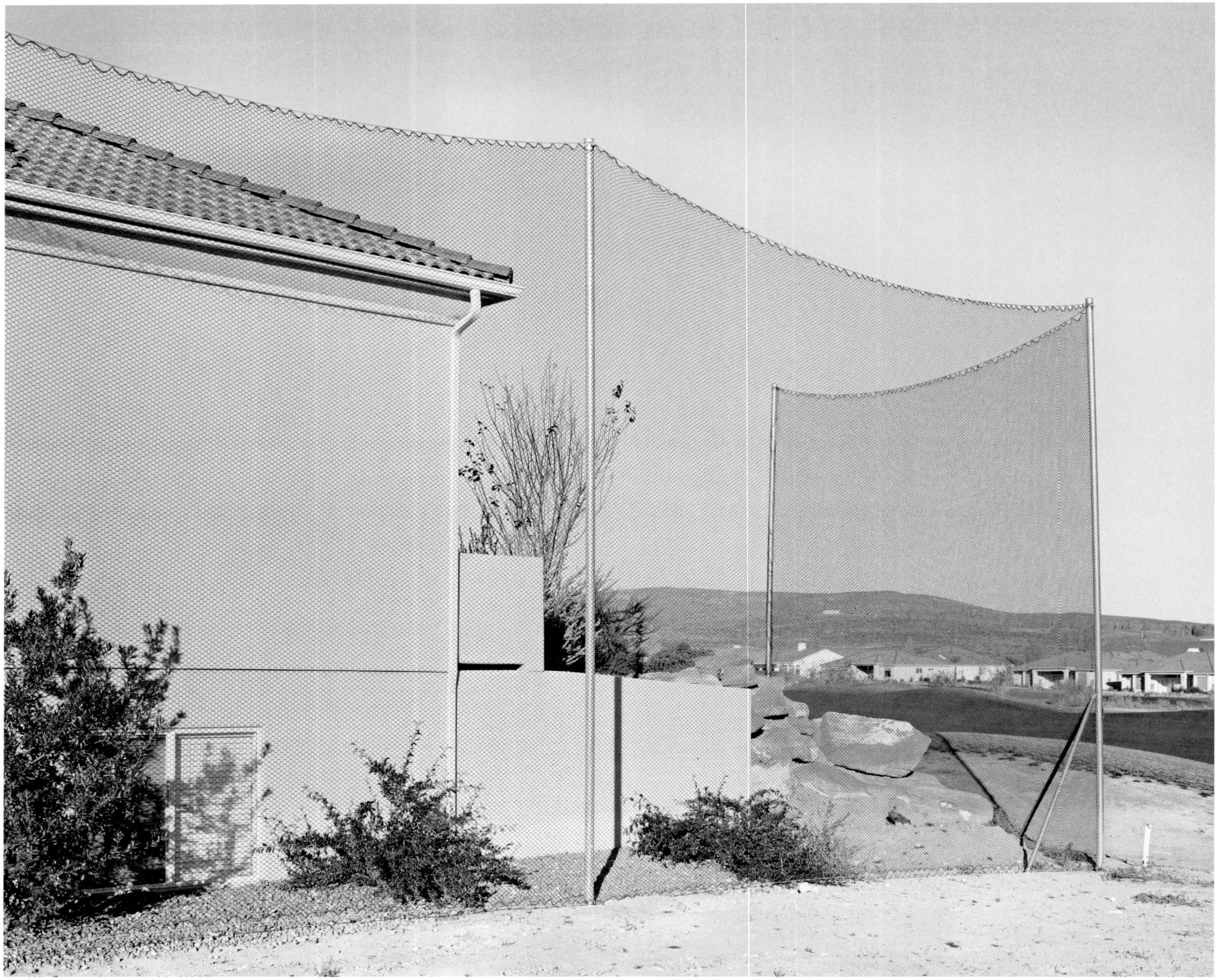

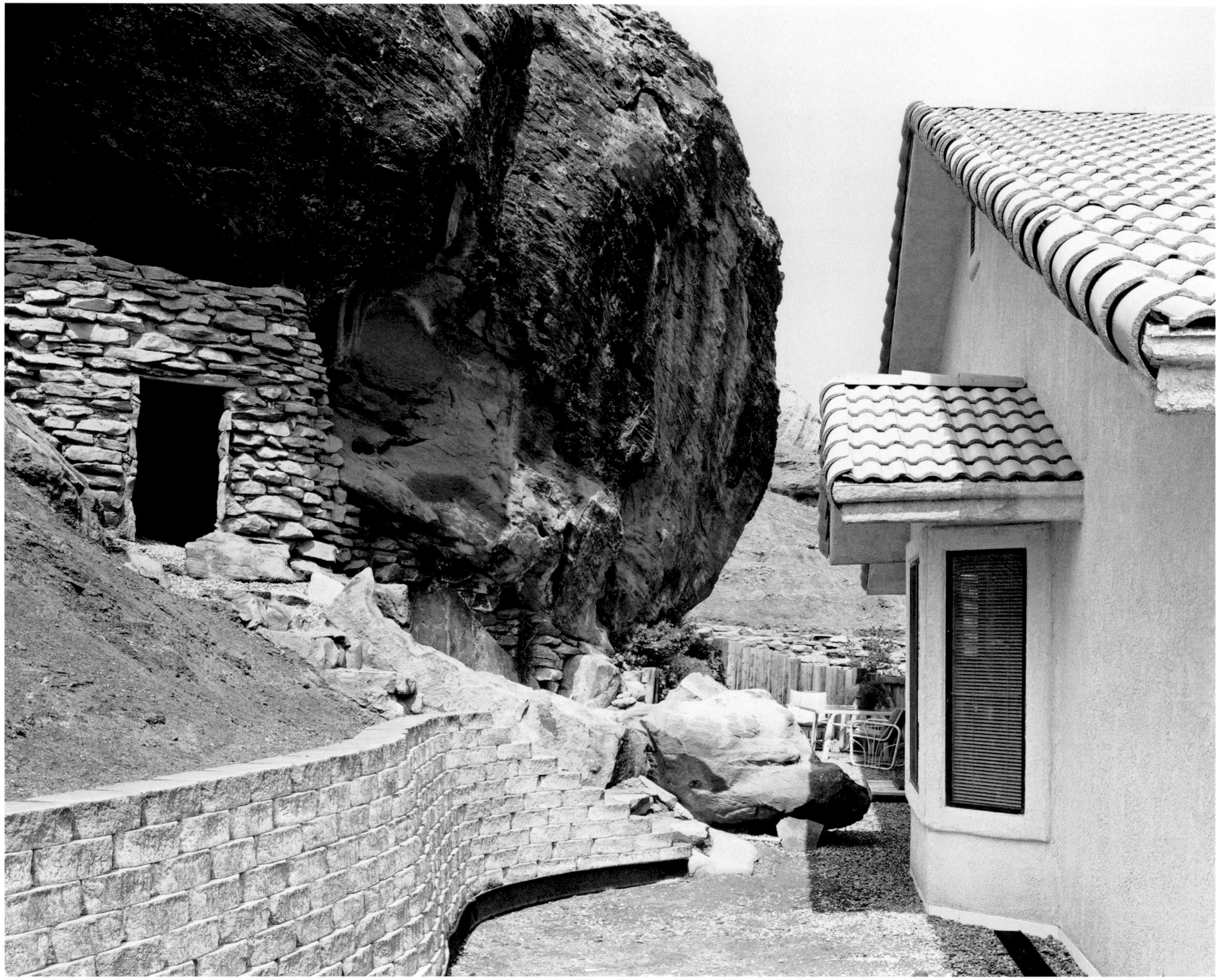

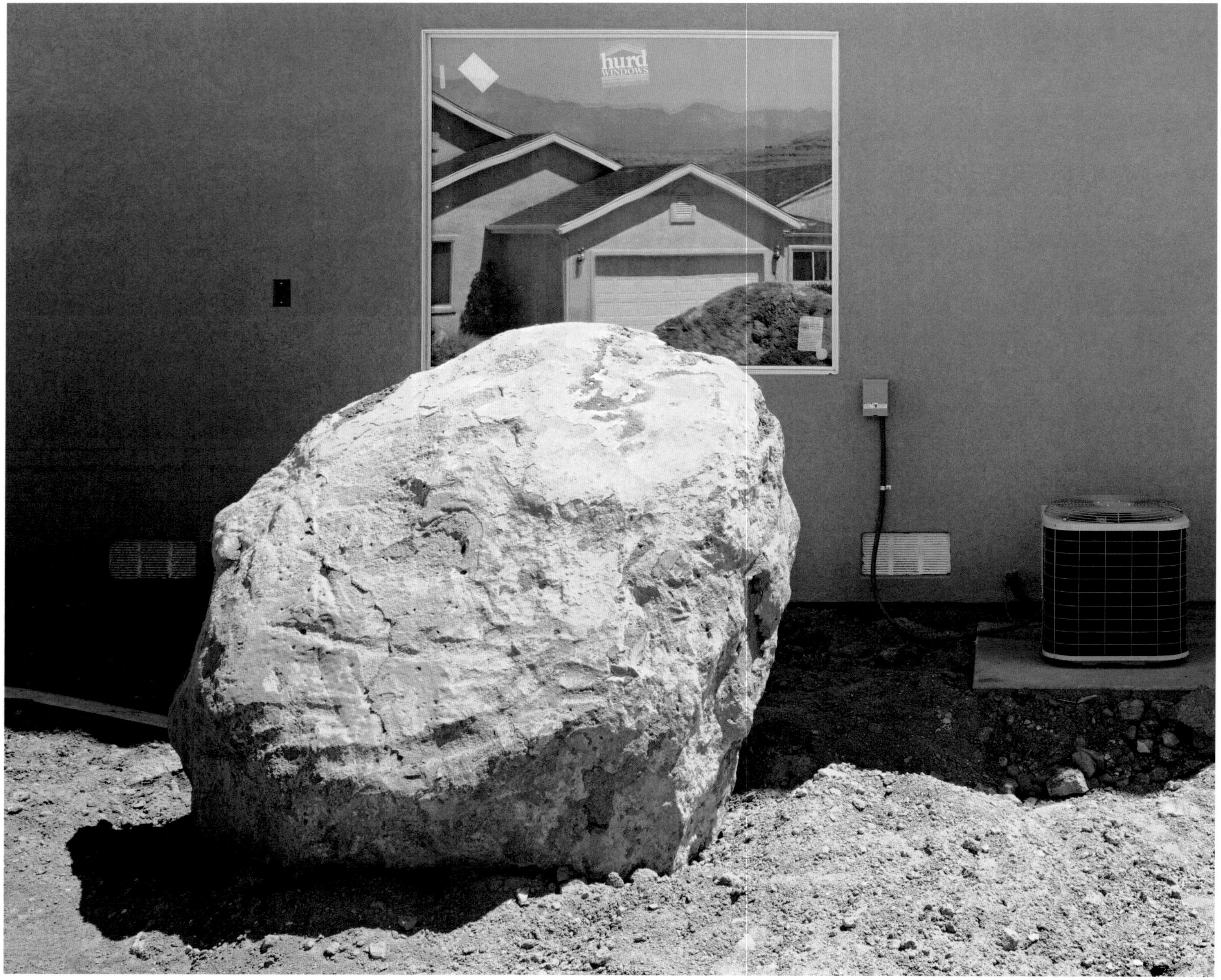
hurd
WINDOWS

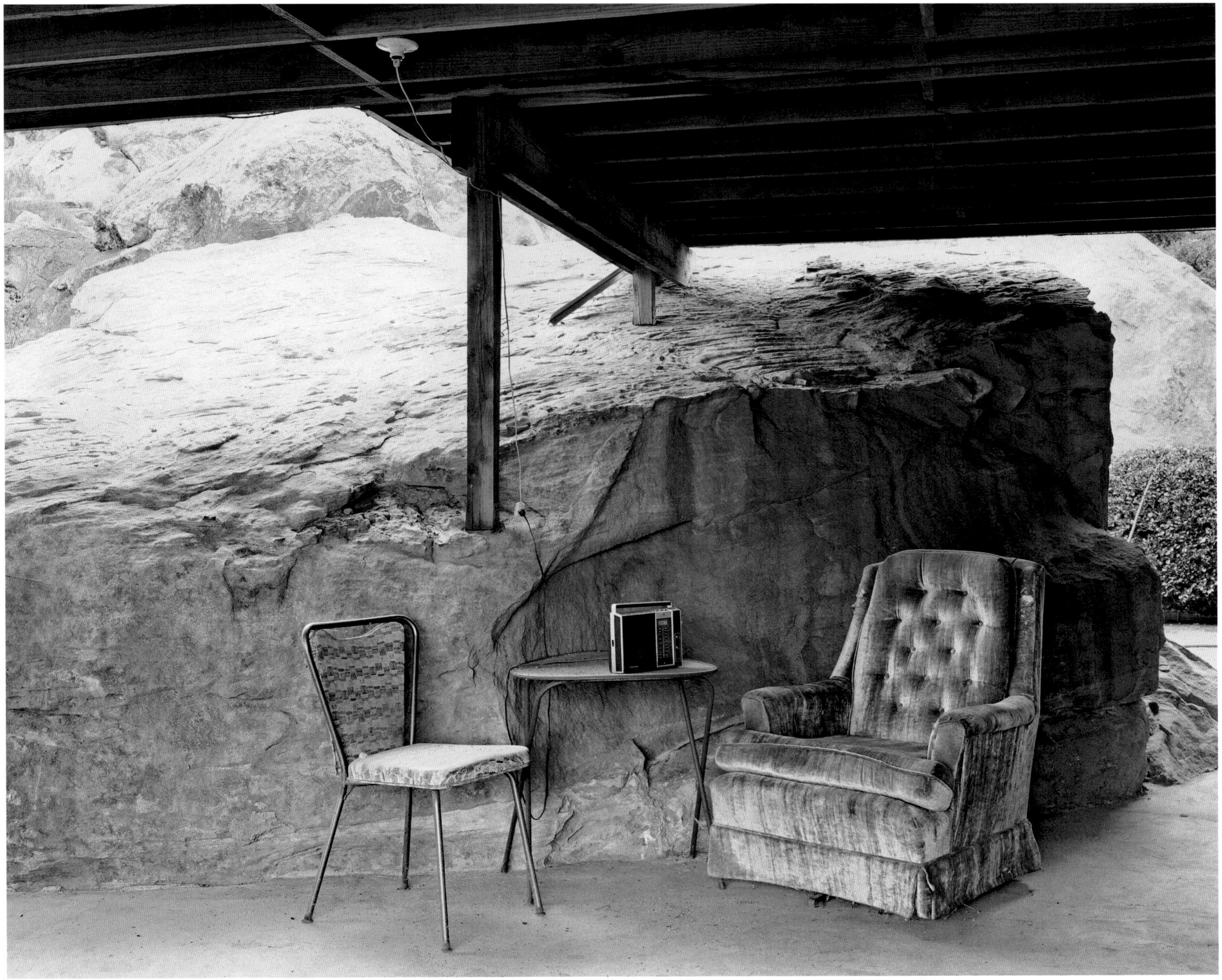

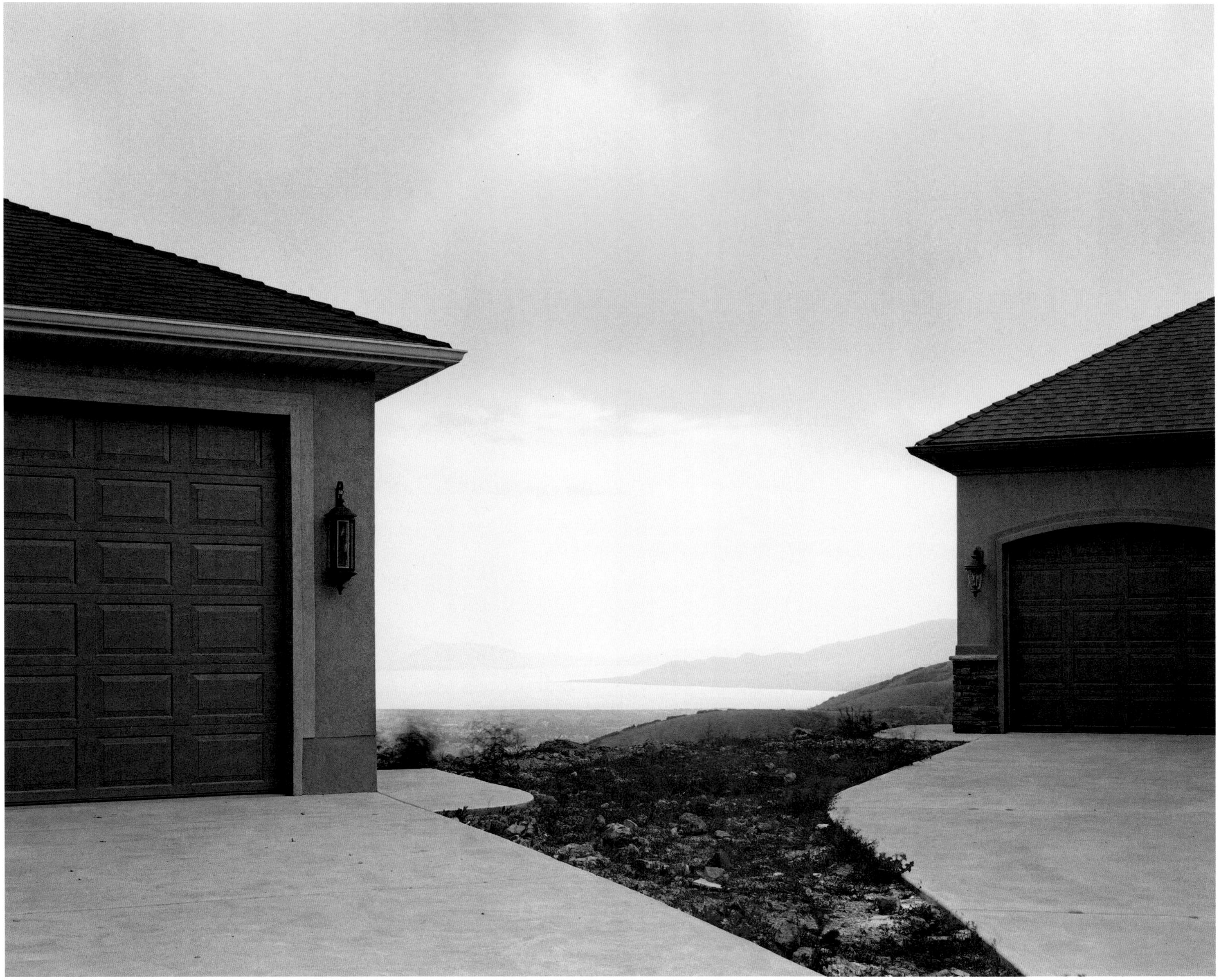

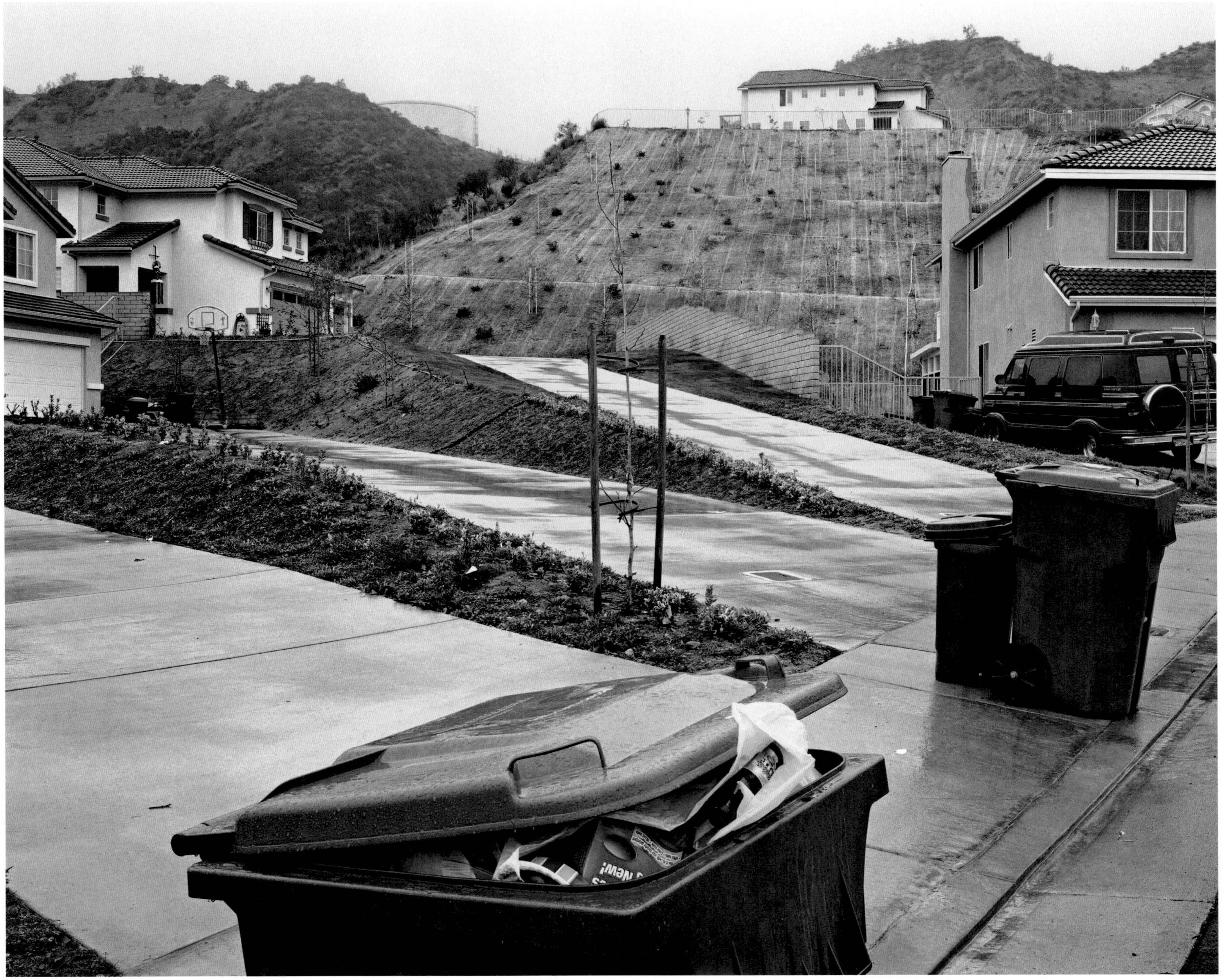

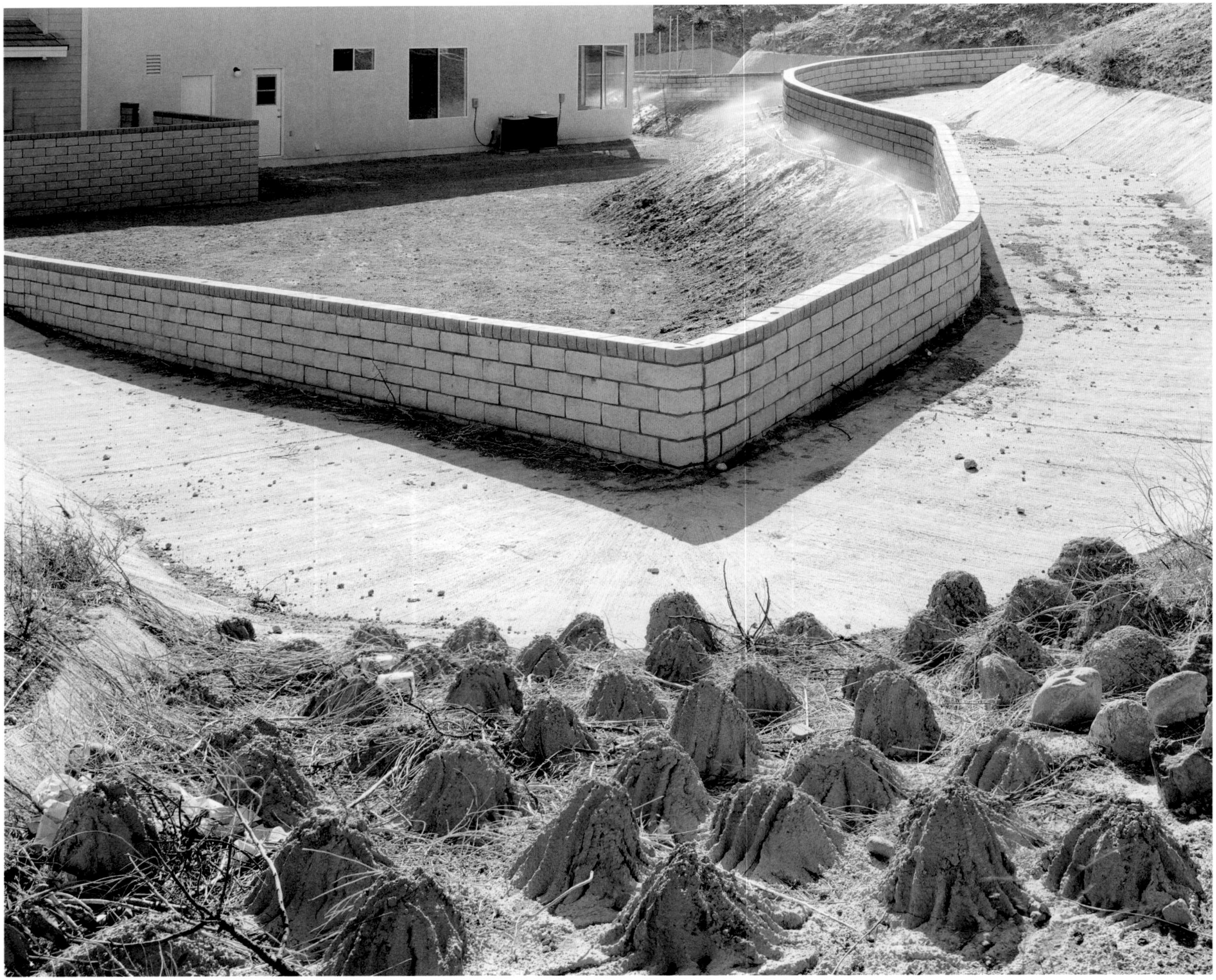

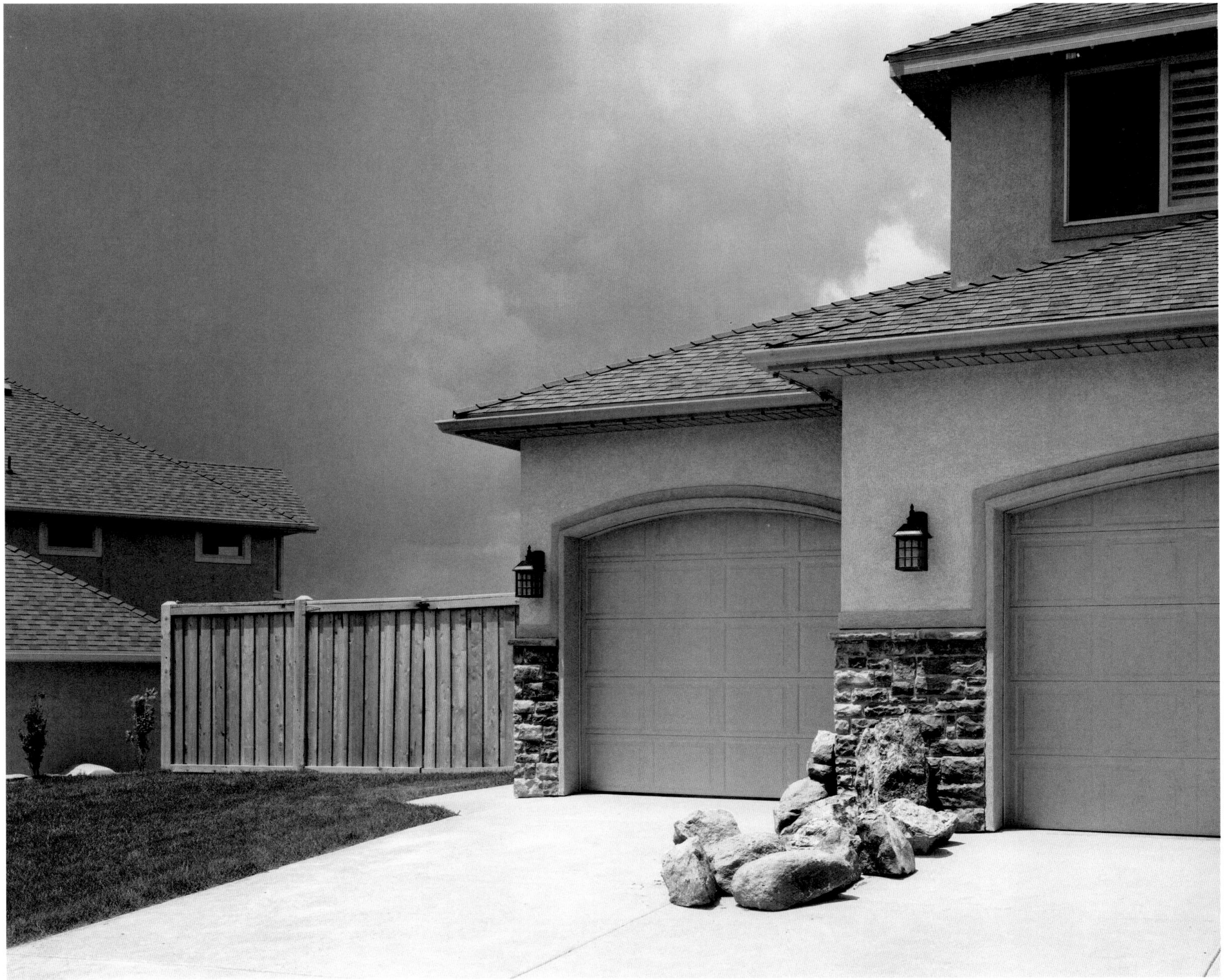

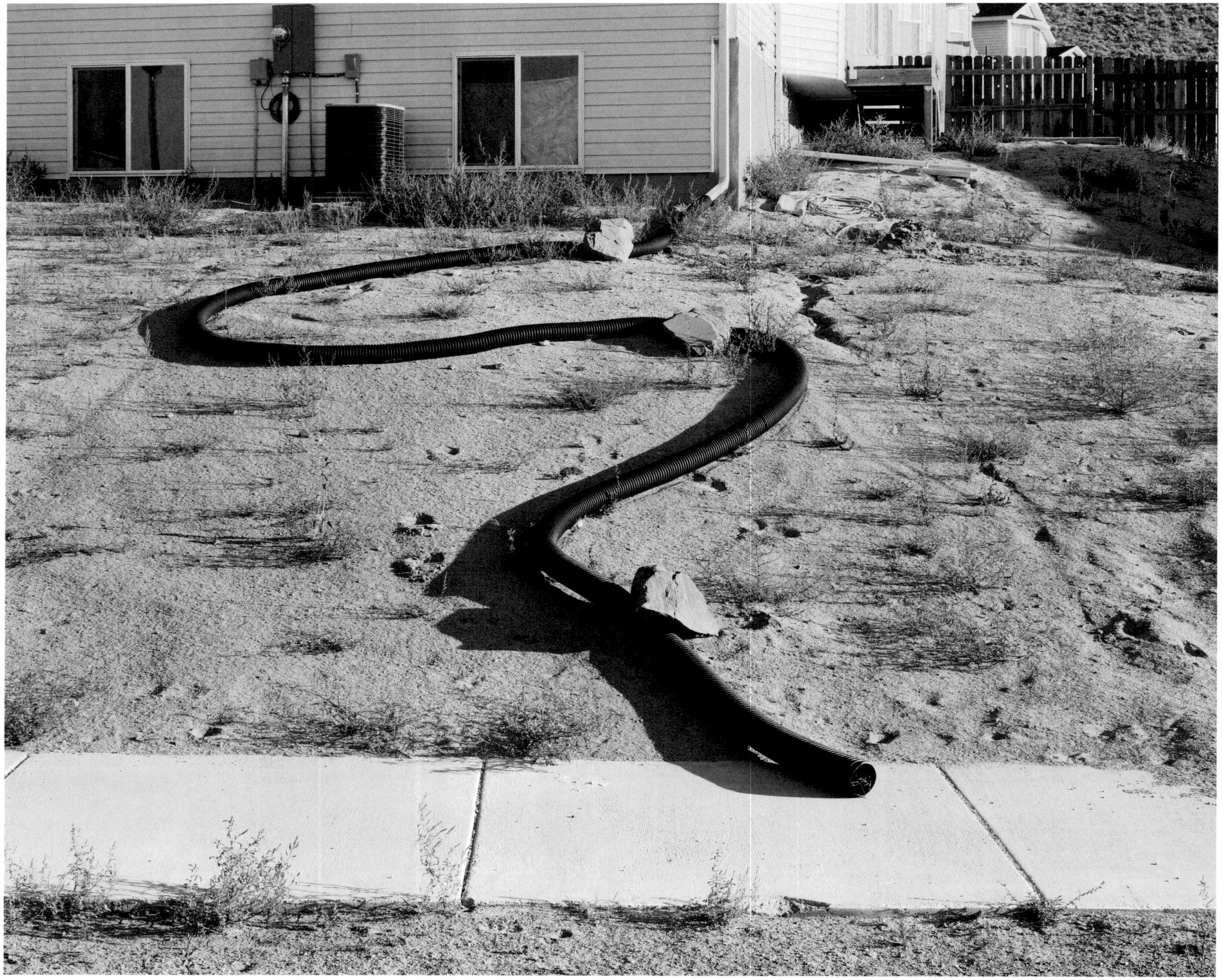

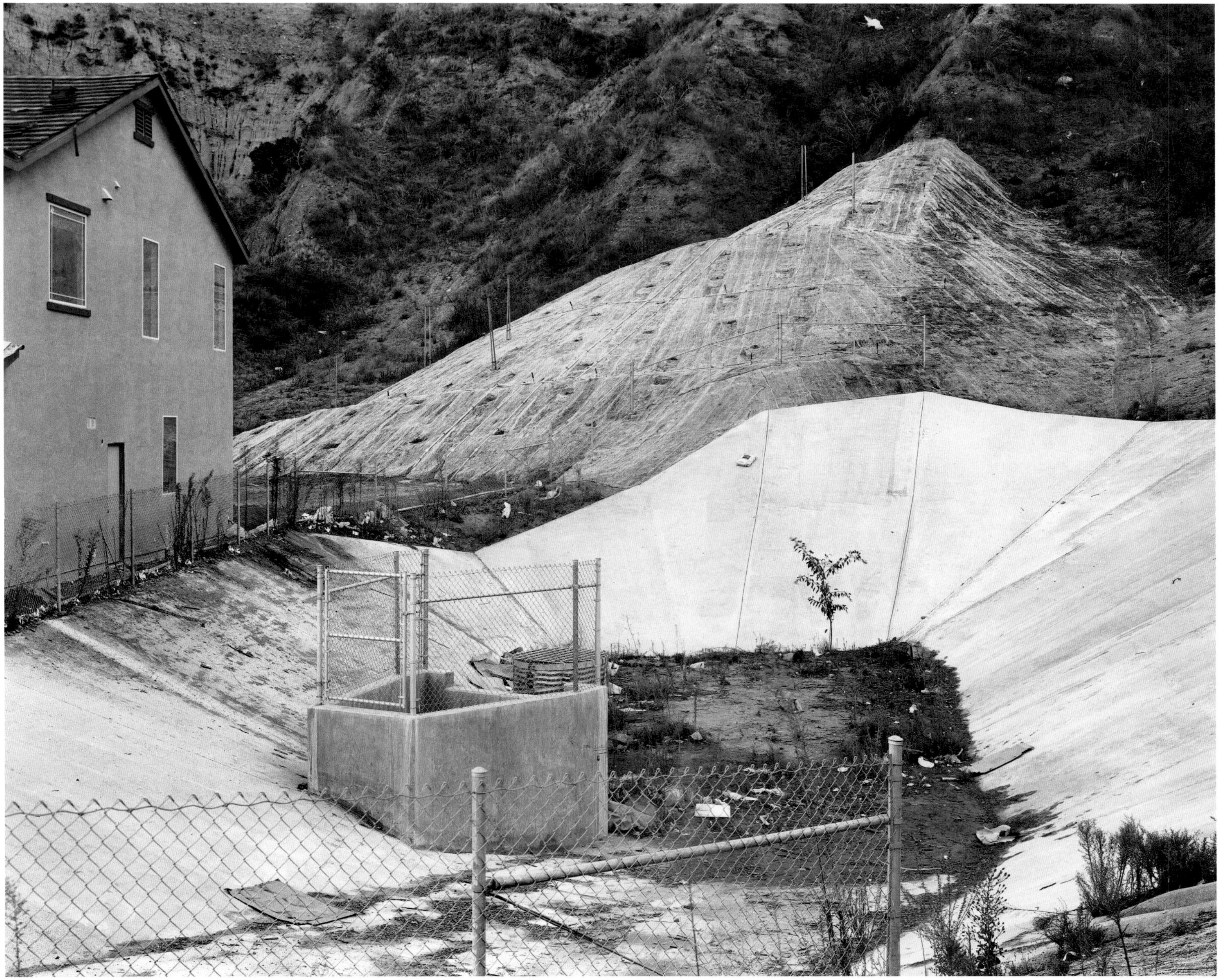

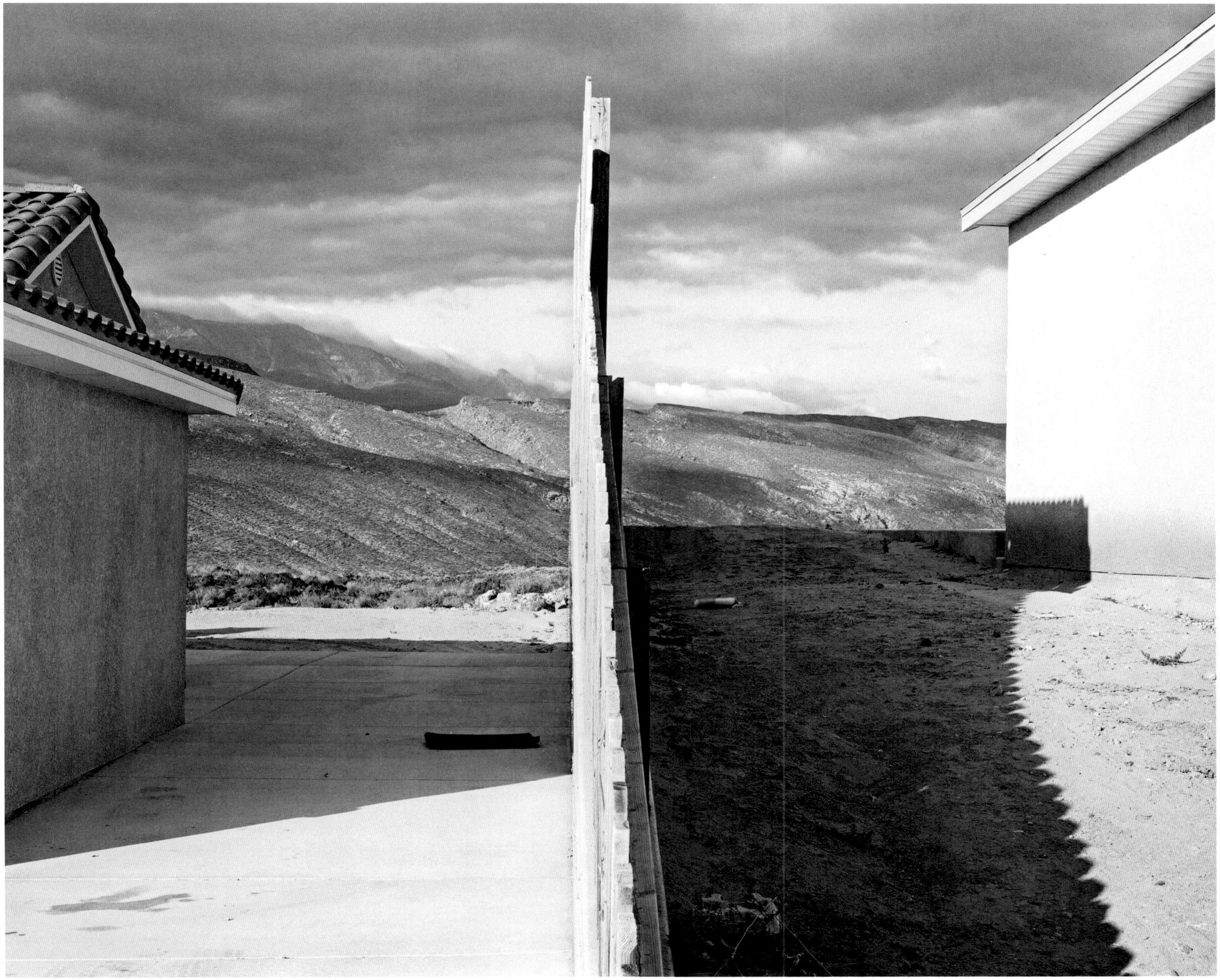

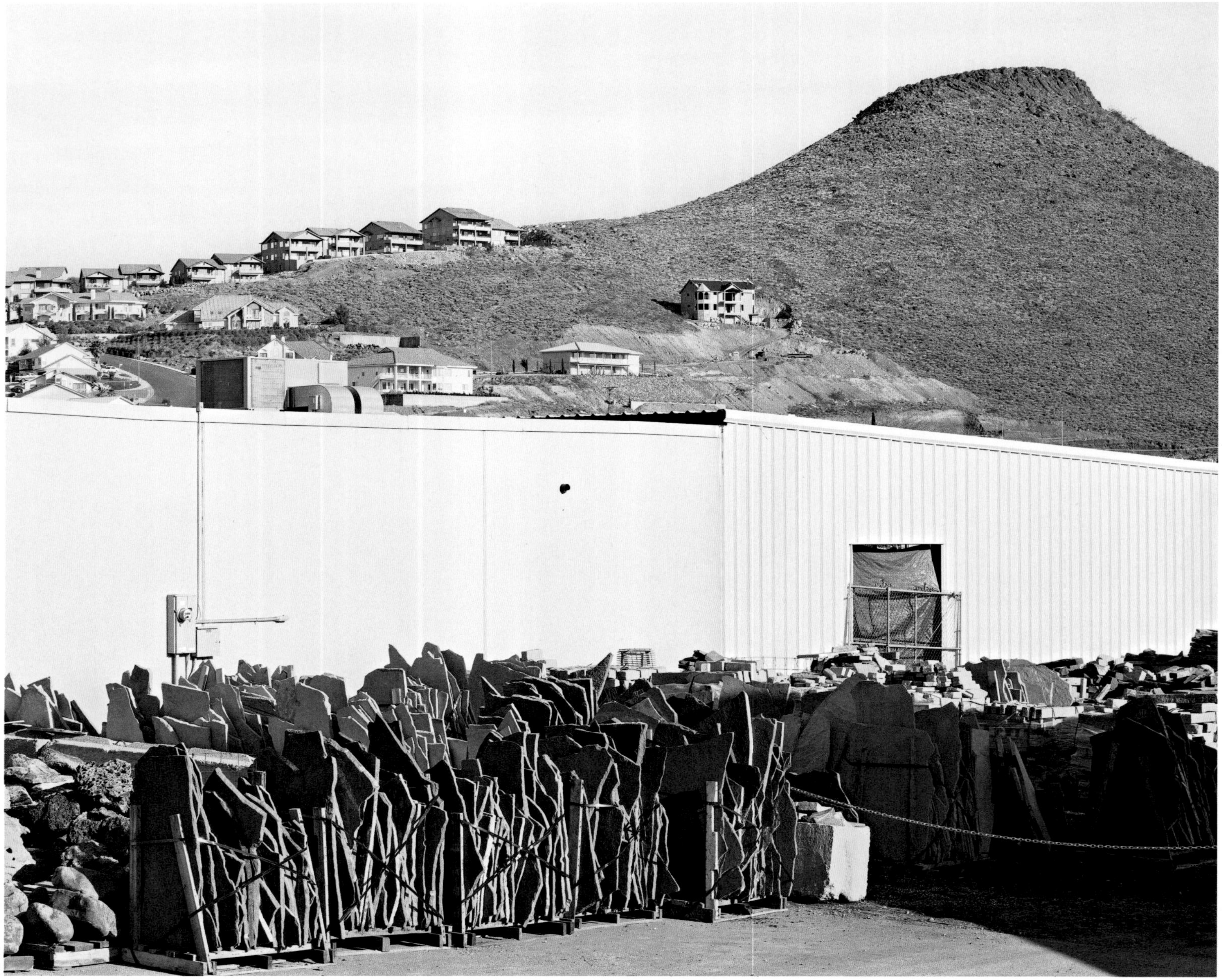

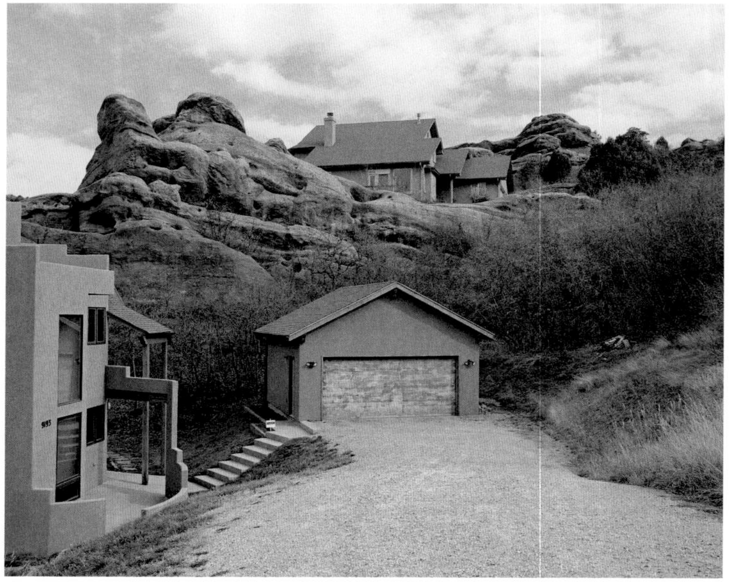
9193

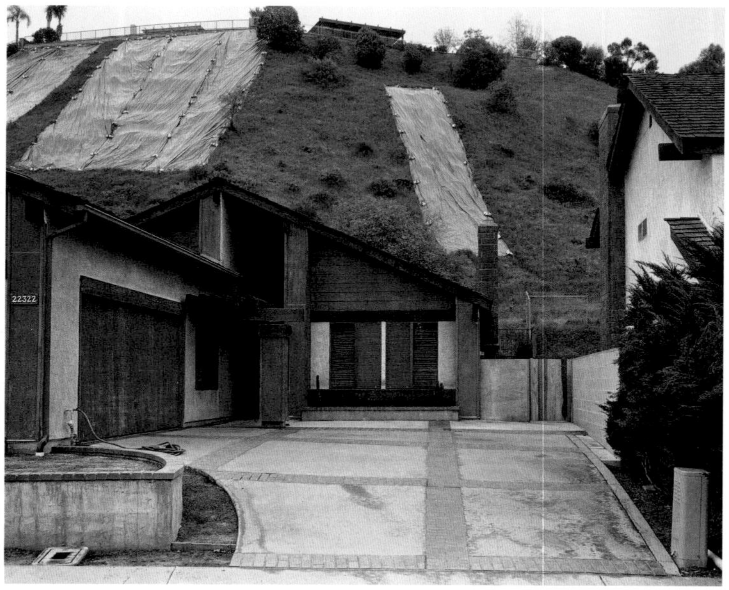
22322

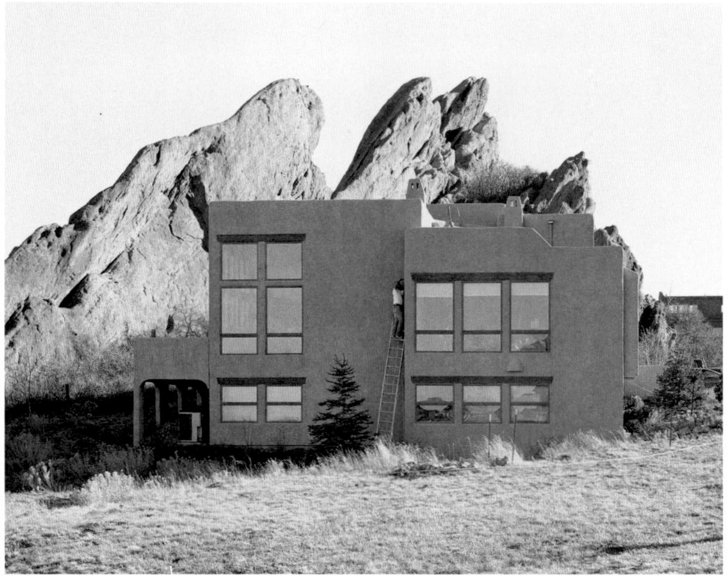

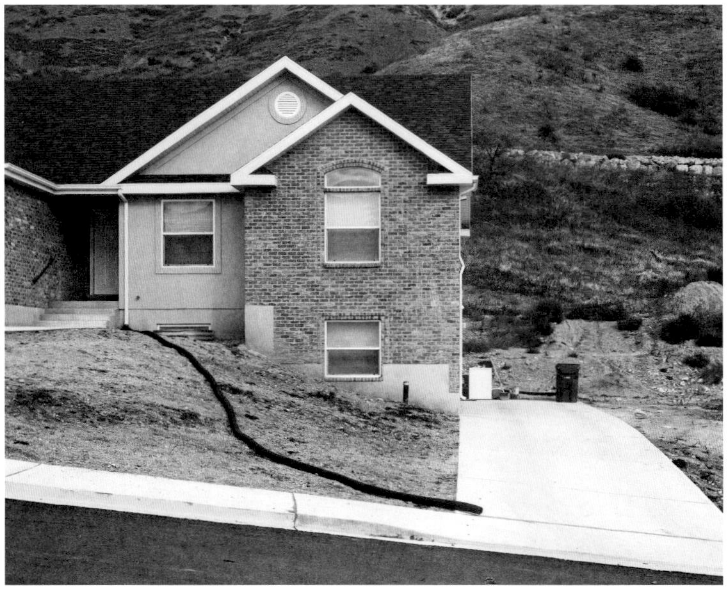

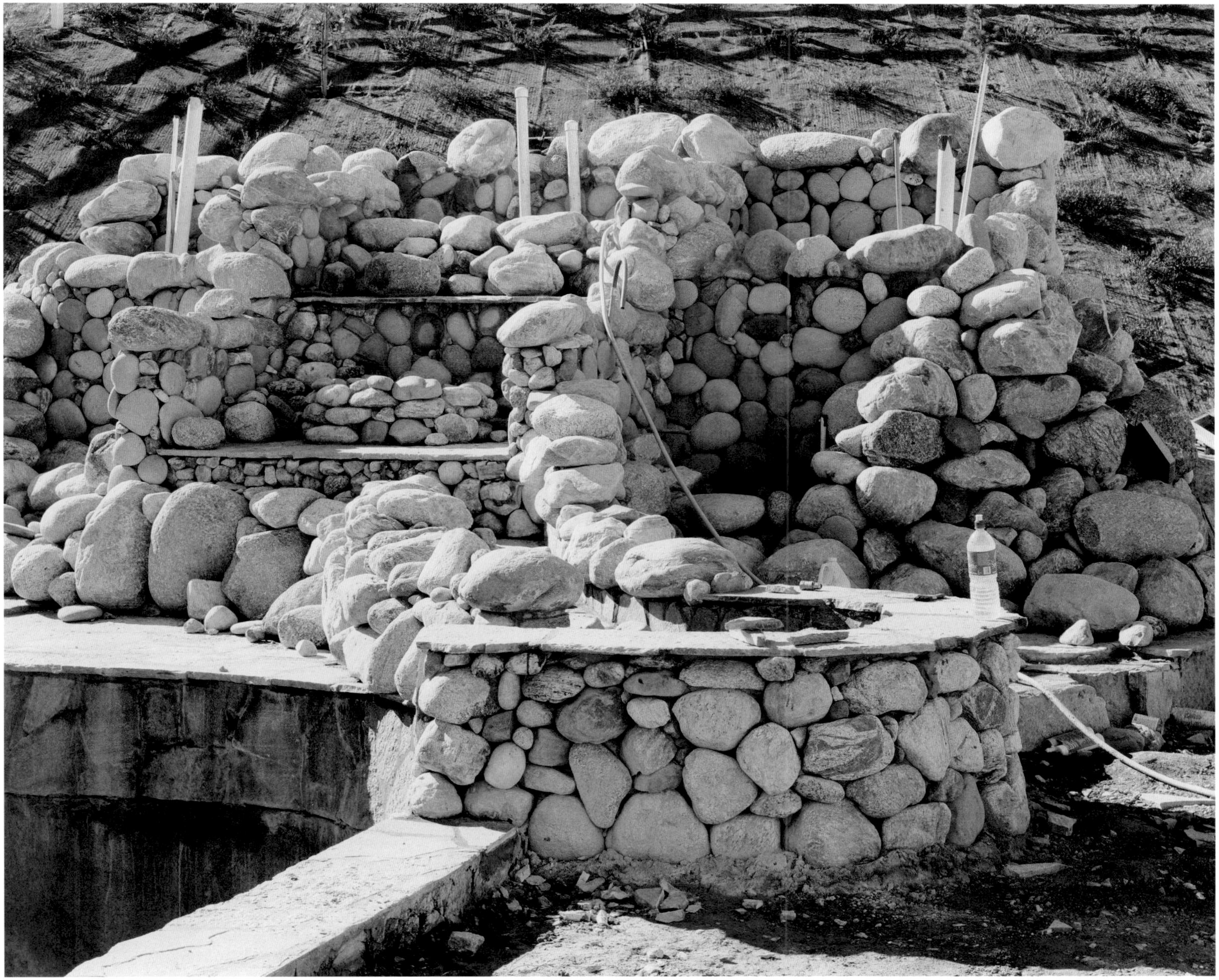

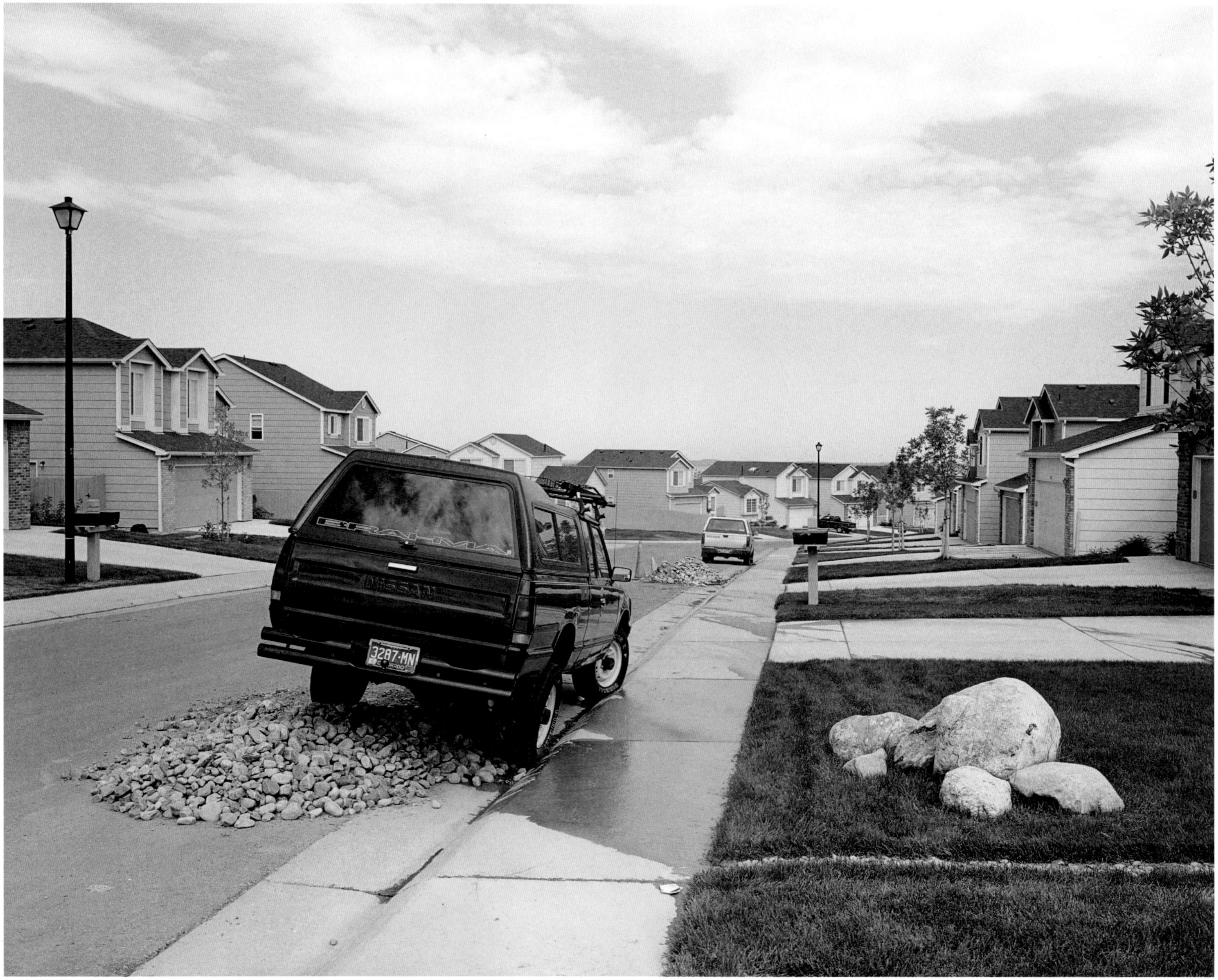
NISSAN
3287·MN

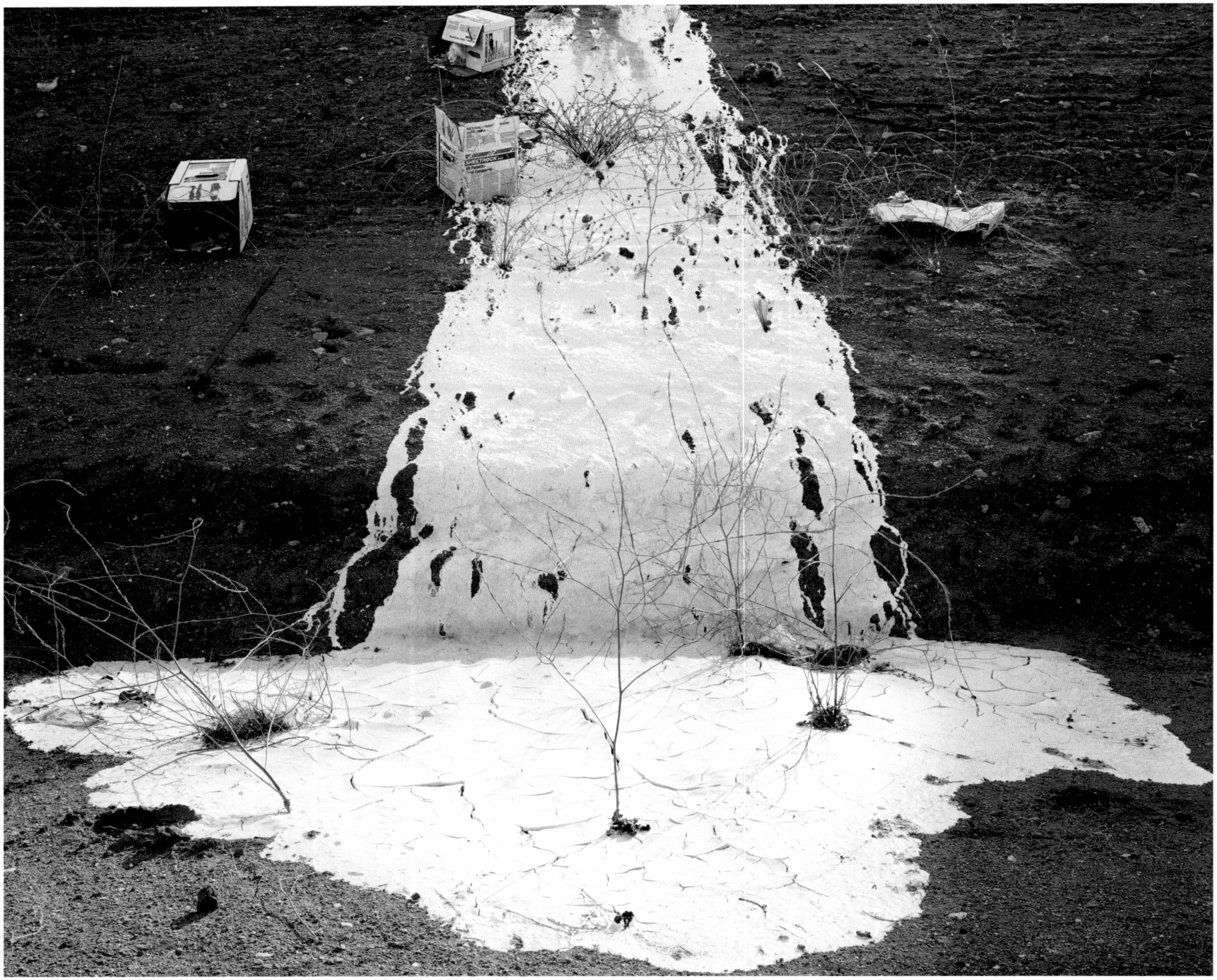
SHEETROCK

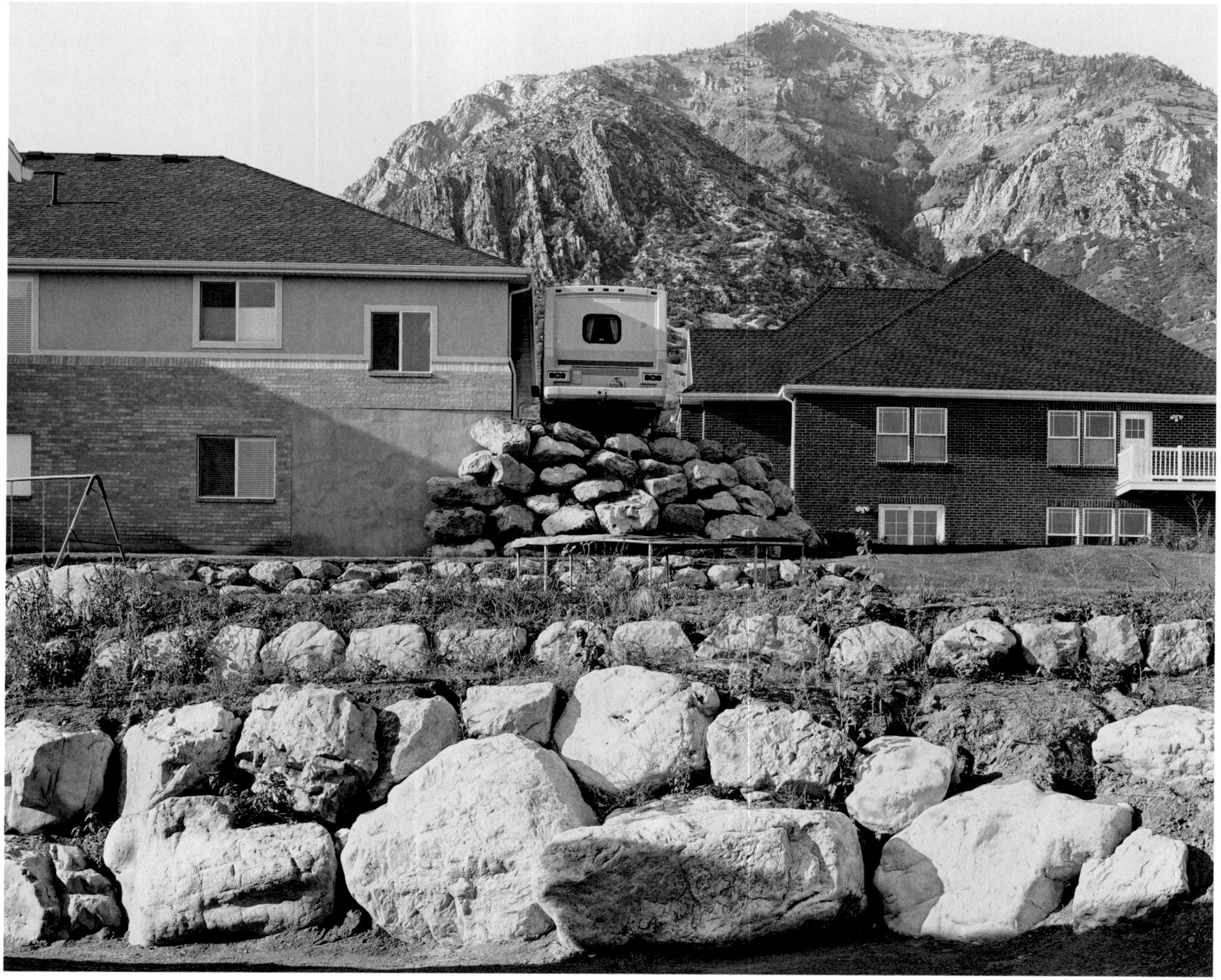

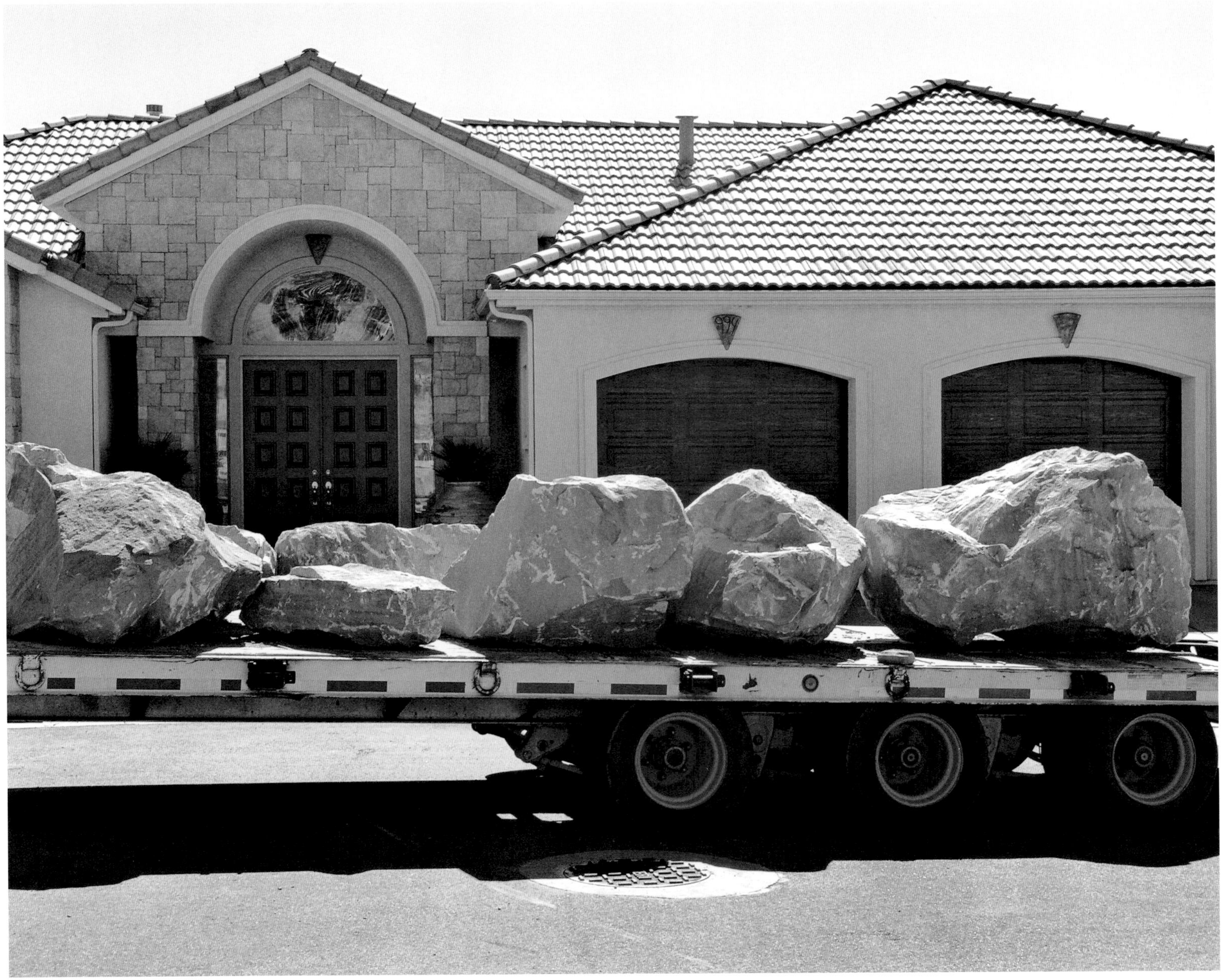

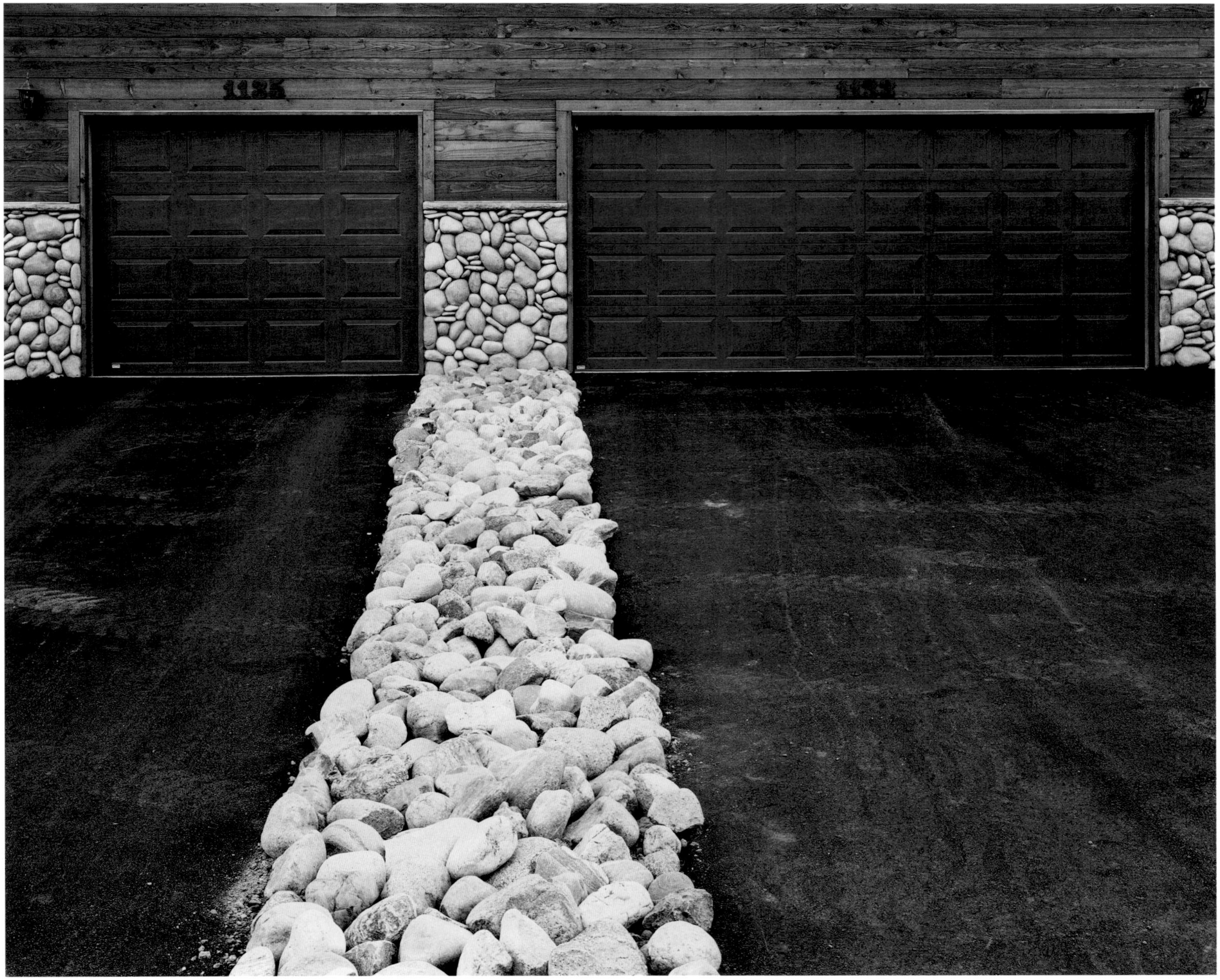
1135

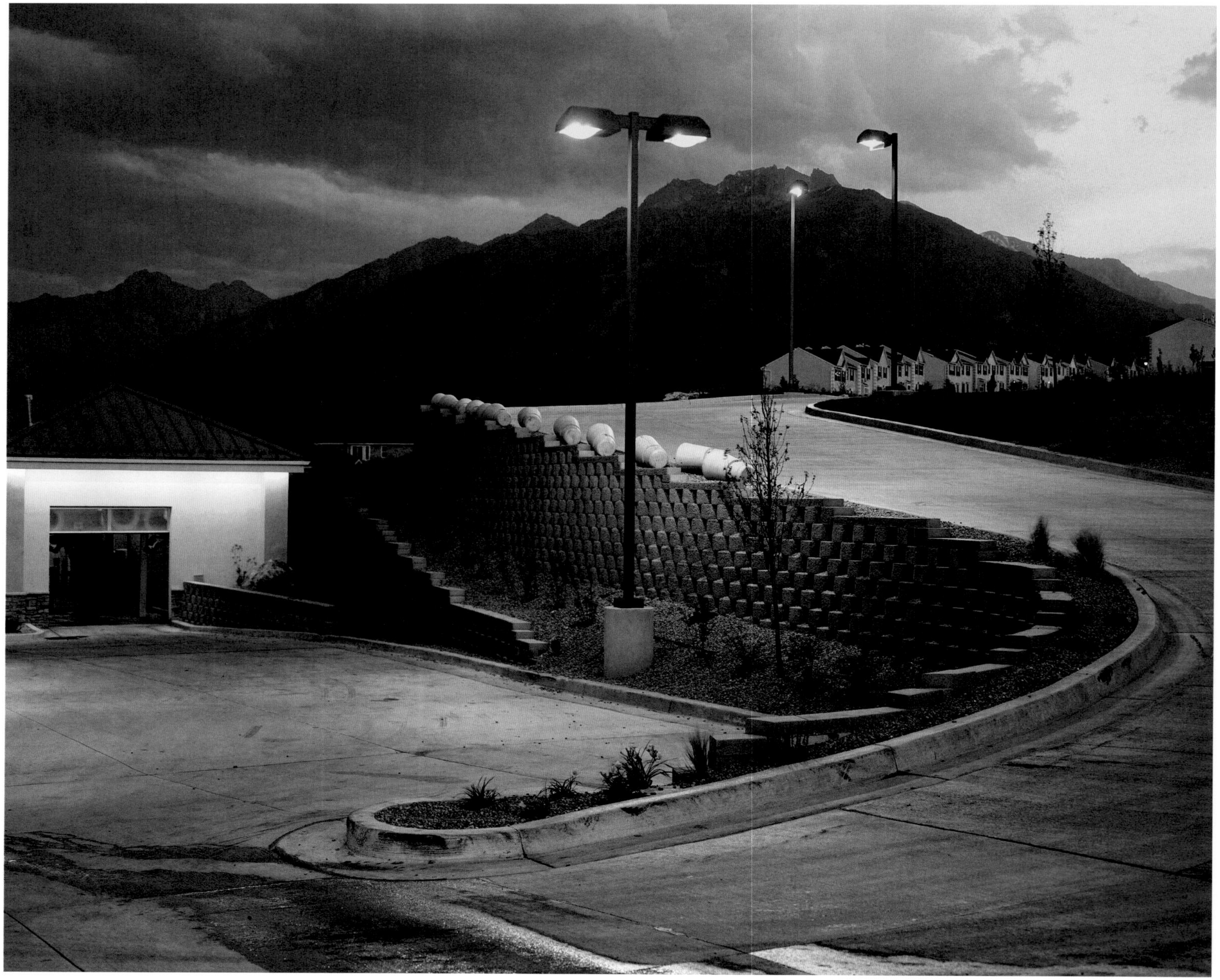

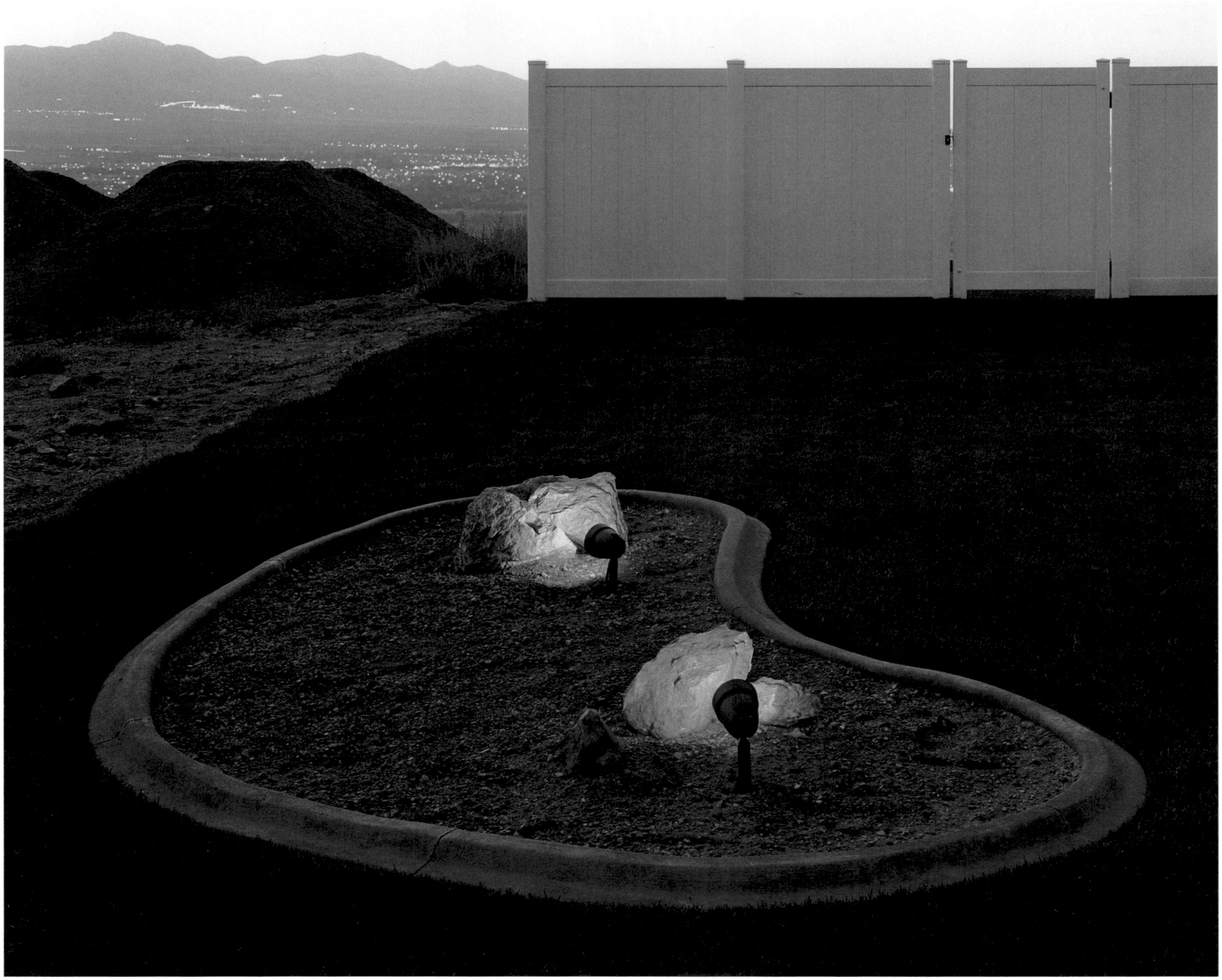

# INDEX

# ACKNOWLEDGMENTS

I would like to express my gratitude to Lynne Honickman at the Honickman Foundation without whom this amazing opportunity to publish my book would not exist. Also a very special thanks to Iris Tillman Hill and Alexa Dilworth at the Center for Documentary Studies at Duke University for all of their patience and expertise, and to the wonderful staff at Duke University Press. I would also like to thank Maria Morris Hambourg for her integrity and insight, and DJ Stout and Julie Savasky for designing such a beautiful book. It is impossible to thank all of those who have contributed to this project. I have started this list, but I feel that it is one I can never finish. It has been a great pleasure to share my life and passion with all of you.

Craig Law
Richard Benson
Dorothy D'Amico
Anne Fessler
Henry Horenstein
Joel and Sandy Barkan
Joe and Alison Crayhon
David Rossi
Ben Smith
Ralph Smith
Don Lyle
Peter Briggs
Tod Papageorge
Justin Kimball
Richard Woodward
Bronwyn Barkan
Victoria Crayhon
Richard Fishman
Rachel Stevens
Mike Smith
Fred and Bonnie Sorenson
JoAnn Walters

The John Simon Guggenheim Foundation
Rhode Island State Council for the Arts
Rhode Island School of Design
Brown University Visual Arts Department

# ABOUT THE PRIZE

The Center for Documentary Studies/Honickman
First Book Prize in Photography

This biennial prize offers publication of a book of photography, a $3,000 award, and inclusion in a traveling exhibition of prizewinners. Each competition a distinguished individual in the field of photography is chosen to judge the prize and write an introduction to the winning book. The prize is open to American photographers who use their cameras for creative exploration, whether it be of places, people, or communities; of the natural or social world; of beauty at large or the lack of it; of objective or subjective realities. The prize will honor work that is visually compelling, that bears witness, and that has integrity of purpose. *The Weather and a Place to Live: Photographs of the Suburban West* is the second book in the series.

Robert Adams chose Larry Schwarm to win the inaugural prize competition for his series of color images capturing dramatic prairie fires. Schwarm's book is titled *On Fire.*

Designed by DJ Stout and Julie Savasky, Pentagram Design, Austin, Texas

The paper for this book meets the guidelines for permanence and durability of the Committee on Production Guidelines for Book Longevity of the Council on Library Resources.

Manufactured at The Stinehour Press, Lunenburg, Vermont
Cover photograph: Backyard, Valencia, California, 1996

Library of Congress Cataloging-in-Publication Data
Smith, Steven B.
The weather and a place to live : photographs of the suburban west / Steven B. Smith.
p. cm.
Includes index.
ISBN 0-8223-3611-1 (cloth : alk. paper)
1. Documentary photography–West (U.S.) 2. Landscape photography–West (U.S.)
3. Suburban homes–West (U.S.)–Pictorial works. 4. House construction–West (U.S.)–Pictorial works.
5. Suburban homes in art. 6. Smith, Steven B. I. Title.
TR820.5.S5393 2005
779'.930774'0978–dc22
2005011685

Duke University Press
Box 90660
Durham, North Carolina 27708-0660
www.dukeupress.edu

CDS Books, published by the Center for Documentary Studies at Duke University, are works of creative exploration by writers and photographers who convey new ways of seeing and understanding human experience in all its diversity – books that tell stories, challenge our assumptions, awaken our social conscience, and connect life, learning, and art. These publications have been made possible by the generous support of the Lyndhurst Foundation.

Center for Documentary Studies at Duke University
http://cds.aas.duke.edu

First printing